GOLD RUSH STORIES

GOLD RUSH STORIES

49 TALES OF SEEKERS, SCOUNDRELS, LOSS, AND LUCK

GARY NOY

Foreword by Gary F. Kurutz

Heyday, Berkeley, California
Sierra College Press, Rocklin, California

This Sierra College Press book was published by Heyday and Sierra College.

Library of Congress Cataloging-in-Publication Data
Names: Noy, Gary, 1951- author.
Title: Gold rush stories : 49 tales of seekers, scoundrels, loss, and luck / Gary Noy ; foreword by Gary F. Kurutz.
Description: Berkeley, California : Heyday ; Rocklin, California : Sierra College Press, [2017] | Includes bibliographical references.
Identifiers: LCCN 2016048917| ISBN 9781597143844 (pbk. : alk. paper) | ISBN 9781597143868 (Amazon Kindle)
Subjects: LCSH: California--Gold discoveries--Anecdotes. | California--History--1846-1850--Anecdotes. | California--History--1846-1850--Biography. | Gold mines and mining--California--History--19th century--Anecdotes. | Gold miners--California--Biography. | Frontier and pioneer life--California--Anecdotes.
Classification: LCC F865 .N68 2017 | DDC 979.4/04--dc23
LC record available at https://lccn.loc.gov/2016048917

Cover Art and Design and Interior Design/Typesetting: Ashley Ingram
Printing and Binding: Printed in East Peoria, IL, by Versa Press, Inc.

Gold Rush Stories was published by Heyday and Sierra College.
Orders, inquiries, and correspondence should be addressed to:
Heyday
P.O. Box 9145, Berkeley, CA 94709
(510) 549-3564, Fax (510) 549-1889
www.heydaybooks.com

10 9 8 7 6 5 4 3 2 1

FOR MY FATHER AND GRANDFATHER,
CORNISH GOLD MINERS

CONTENTS

FOREWORD

GARY F. KURUTZ

When James Wilson Marshall's eye was "caught with the glimpse of something shining in the bottom of the ditch" at Coloma more than 165 years ago, he not only set in motion a worldwide rush to California but also touched off the greatest writing frenzy in our nation's history. That floodtide of books and articles concerning the Golden State's defining event continues to this day, but, until *Gold Rush Stories*, never has there been a publication that so wonderfully encapsulates the real and mythological, the triumphs and tragedies, and the picturesque and mundane of that wild and rambunctious era. Gary Noy's book is both factual and entertaining, and his dynamically told narrative presents a splendid overview of the California Gold Rush from its beginnings until full maturation.

Noy, a native of the captivating and historic Gold Rush town of Grass Valley, has become the master storyteller of the High Sierra and, perforce, of the Mother Lode. After all, the precious metal that brought the world to California was found in Sierra streams and hillsides. Given his heritage, it is fitting, then, that Noy has assembled this treasure trove of forty-nine stories. Moreover, his grandfather and father both worked at the renowned Empire Mine, and there is no doubt that they, too, both imparted dramatic accounts of working at this famous hard rock gold mine. Noy grew up within earshot of the giant stamp mills and likely witnessed with boyish wonder the turning of giant Pelton waterwheels and massive trucks rumbling through town or the dozens of dust-covered workmen ascending from a mine shaft during a shift change. The narrow and irregular streets of his Grass Valley, on fabled Route 49, are redolent with history. The hotels, bars, and churches bore witness to an ongoing pageant. Stories about legendary Grass Valley figures such as California's first humorist, Alonzo Delano, and John Rollin Ridge, the Native American author of *The Life and Adventures of Joaquín Murieta*, form a vital part of Noy's "warp and woof."

Consequently, it is not surprising that this son of Grass Valley has gone on to a distinguished career as a historian and teacher with a focus on his beloved Sierra Nevada mountain range and the Gold Rush. For many years, Noy has taught history at Sierra Community College in Rocklin, Placer County. How fortunate his students have been to hear his lectures and be inspired by his stories. Brimming with knowledge and enthusiasm, Noy has successfully taken the history of this vast region and its intellectual treasures beyond the classroom through several different ventures. From 2002 to 2012, he served as the founder and director of SCC's Center for Sierra Nevada Studies—the type of program more readily found at a university with a doctoral program. As part of the Center, and taking advantage of emerging technology, he created a virtual museum of Sierra history, a fabulous way to bring its rich past to a broad audience online. Showing even more ambition, Noy directed the Sierra College Press as editor-in-chief and developed a strong partnership with the highly acclaimed regional publisher Heyday. Again, such an endeavor is with this one exception invariably centered at a major university. Recognizing the vital importance of historical documentation, this perspicacious pedagogue also established the Archives of Sierra Community College History. Through Gary Noy's energetic and imaginative leadership, he has created a historical, literary, and artistic program as big as the college's geographic namesake.

With the passion of an evangelist, Noy has given more than three hundred talks to community and academic groups, bringing to diverse audiences the wonderful and picturesque stories of this beautiful region. Because of these speaking engagements and passion for regional history, Noy knows Gold Country from Quincy to Mariposa better than anyone. He has that wonderful ability to bring out the memorable and unusual that will equally pique the interest of the serious academic and the inquisitive resident who simply wants to learn more about his or her community. This gentle scholar brings to life the grizzled forty-niner, the beguiling dancehall girl, the scolding fundamentalist preacher, the nattily attired card shark, the industrious Chinese miner squatting at a stream with his wash pan, and the enterprising woman who baked pies for homesick miners. Fortunately, too, this Sierra sage has put these magnetic stories into print over the years. Through that wonderful partnership with Heyday, he has published *Sierra Stories: Tales of Schemers, Dreamers, Bigots, and Rogues* (2014) and *The Illuminated Landscape: A Sierra Nevada Anthology* (2010). *Gold Rush Stories*, then, is a natural sequel to these two highly acclaimed books.

The rich body of eyewitness accounts and recollections found by Noy in rare books, periodicals, newspapers, and previously unpublished letters forms the foundation of this rich tapestry of tales. Many of the leading contemporary authors are represented here—among them Bayard Taylor, Alonzo Delano, John David Borthwick, and Eliza Farnham—but our author also mined deeper and deeper to find previously unpublished accounts in historical repositories, lifting from the tiny six-point type of era newspapers dramatic eyewitness accounts deserving of recognition.

One of the more familiar names is James Mason Hutchings, most famous as the founder and editor of *Hutchings' Illustrated California Magazine*, but also the author of California's first bestseller, "The Miners' Ten Commandments." Noy recounts how Hutchings was the first to really capitalize on the tourist appeal of the Yosemite Valley, establishing a hotel there and hiring one of the most sainted names in California history—John Muir—to run his sawmill, all while his own eminence as a writer and publisher continued to grow. He, more than anyone, saw the potential of a distinctly California innovation, the pictorial letter sheet, one side of which featured an engraving or lithograph of a California scene, with blank space on the back for letter writing—a foreshadowing of the picture postcard. His work as both a businessman and a writer influenced generations to come, including a young Ansel Adams, whose love of nature was in part inspired by Hutchings's classic *The Heart of the Sierras*.

Even as many of the personalities and events featured herein are familiar to students of this golden stampede, Noy presents them in a context that a twenty-first-century reader will especially appreciate. He is a gifted writer whose words enable those delving into these rich pages to clearly visualize, taste, and smell the events he has selected. In creating this compendium, Noy is sensitive, too, to the unparalleled diversity of the region. Technically, California was still a Mexican province when Marshall made his momentous discovery, and this remote locale featured a population made up not only of the mixture of American and European settlers who came to the region during the rush but also of Native Americans and Californios of Hispanic heritage who had been here for generations. The tragic plight of the Indians and dispossession of the Californios are well represented in these chapters. Also among those who suffered much during this era were the recently arrived Mexicans skilled in mining techniques, the Chinese workers seeking their own "Gum Shan," or "Gold Mountain," the Jewish merchants seeking business opportunities, and the thousands of Frenchmen escaping the political tumult of their own country. African Americans came to California either as slaves or freedmen, but all hoped to find not only their "pile" but also acceptance.

Keep in mind that the Gold Rush was, after all, our nation's first large-scale nonmilitary encounter with other nations and people. Many American forty-niners had never before been outside their villages or towns, much less a place thousands of miles away. Intrepid New Yorkers, for example, were for the first time encountering the steaming jungles of the Isthmus of Panama and landing at exotic ports in Brazil, Chile, Peru, and Mexico. And when they got to California they met a population diverse in ways they hadn't seen before. As our first governor, Peter Burnett, noted in 1850, "What we have here in our midst is a mixed mass of human beings from every part of the wide earth, of different habits, manners, customs, and opinions, all, however, impelled onward by the same feverish desire of fortune-making."

Dame Shirley, arguably the most eloquent of all Gold Rush writers, summed it up best in one of her 1852 letters published in California's first magazine, the *Pioneer*:

> You will hear in the same day the lofty melody of the Spanish language, the piquant polish of the French . . . the silver, changing clearness of the Italian, the harsh gargle of the German, the hissing precision of the English, [and] the liquid sweetness of the Kanaka.

The Gold Rush was also kind to many individuals, although perhaps not as kind in the long term as would be expected in some particular cases. Despite having the advantage of location and influence, two of the key figures central to the great discovery—James Marshall and Captain John A. Sutter—did not emerge from the era wealthy and content. Marshall, who had under Sutter's direction helped build a sawmill at Coloma, would lead a troublesome and unhappy life, hounded by people thinking he had magical powers to detect the precious metal. Sutter's carpenter spent his remaining years frustrated and dying in poverty while Sutter himself, the master of a vast inland empire at the gateway to the goldfields, would see it collapse in short order, overrun by frantic gold seekers and squatters. He would spend the rest of his life trying to receive government compensation but never regained his status as the "Baron of New Helvetia." The adobe walls of his landmark fort would soon melt back into the earth, symbolic of his inability to control events once word of this new Golconda circled the globe.

Through Noy's riveting narrative and stories extrapolated from eyewitness accounts, we also have a profile of the Golden State's tumultuous political history during this pivotal time. With so many fortune seekers and opportunists pouring into the region, the lack of civil government became readily apparent and, in response, vigilante groups lashed and hung transgressors. Still others even thought of forming their own commonwealth if the federal government in Washington, D.C., did not act to bring a semblance of order and admit California into the union. Finally, civil government came to be, but as typical of this maverick state, the first legislature, which established such fundamental laws as prohibiting slavery and giving property rights to married women, ironically became known as the "Legislature of a Thousand Drinks," as seen through the frolicking personality of Senator Thomas Jefferson Green.

Many of the argonauts did find their fortune in California either by digging for nuggets or by "mining the miners"—that is, selling them wash pans, shovels, mules, and food. Still others, especially those beguiling "daughters of joy," did well by merely smiling at a lonesome miner or agreeing to have him buy her a drink or tripping the light fantastic together. Not surprisingly, many of the most memorable accounts from this era are of those roaring "gambling hells" and saloons that took advantage of those who came into town holding sacks of the precious auriferous dust harvested through months of unremitting toil. Naturally, in this rowdy, anything-goes environment, these exhausted gold hunters wanted to let loose by drinking, gambling, dancing, and whoring their way to happiness. Occasionally, they would pack into a theater to listen to and shout over music, sometimes merrily throwing nuggets at actresses who would have been booed off the stage in a Boston venue. Others, wanting more of a release, availed themselves of the services of "good-time girls." The vast majority, of course, wrote about the other guy gambling, getting drunk, or entering into a makeshift bordello. In one extraordinarily rare instance, however, Noy quotes from a previously unpublished and all-too-candid letter from 1853 wherein a miner brags to another about the cheap cost of female services and liberally makes use of "the F word."

In this helter-skelter environment, Gold Country had more than its share of natural and manmade disasters. Sacramento, the great supply center of the interior, was, despite common sense, located at the confluence of two rivers prone to flooding, the Sacramento and the American. In 1850 and 1861–62, the rivers turned the area into a vast inland sea, costing many lives and millions of dollars in damage and lost goods. Other mining camps farther upriver likewise drowned. In other areas, cheaply built wood and cloth structures provided perfect fuel for fires started by careless and out-of-control occupants. San Francisco burnt down seven times in a few short years, as did virtually every settlement of any size. One can only imagine how fires in Grass Valley and Sacramento incinerated precious diaries, letters, and daguerreotypes from that era. In addition to floods and fires, many died as a result of drowning, gunfire (both accidental and not), scurvy, cholera, and a host of other afflictions.

As for progress, getting supplies from the great Pacific Rim port of San Francisco to the mines in the interior gave rise to an incredible transportation system. Within a short time, steamboats plied San Francisco Bay loaded with provisions for Sacramento, Stockton, and Marysville. Speed was of the essence, and overeager captains sometimes fueled their engine boilers to the exploding point, causing many fatalities. Others created a lucrative business by supplying the mining camps via pack mule with provisions ranging from flour to whiskey and the greatest treasure of them all: letters from home. Lonely and anxious miners waited in line for hours and hours for the distribution of mail, and woe to the man who did not receive anything or, worse yet, received news of the death of a loved one back home. Out of this cauldron of chaos came two larger-than-life trailblazers, Cock-Eye Johnson and Snowshoe Thompson, whose stories are told in the pages that follow.

While millions of dollars' worth of gold was harvested, the Gold Rush had many other unintended consequences in the form of environmental destruction. Miners eager to quickly make their fortune and return home did not care if their cutting of millions of trees and the construction of miles-long flumes and makeshift dams would cause erosion and flooding downstream. Others used mercury to better get at the nuggets, unwittingly poisoning the water. As hunting for gold became more industrialized, the region saw the introduction of hydraulic mining, whereby giant monitors or water cannons literally blasted away entire hillsides. However efficient this may have been, it ruined the farms, ranches, and towns downstream and filled the great rivers and streams with silt. After vigorous protests from those affected, the 1884 Sawyer Decision of the Ninth United States District Court put an end to hydraulicking.

Noy is a master of creating a sense of place, and in the preface that follows he takes the fortunate reader to the Mother Lode ghost town of Hornitos in Mariposa County. Drawing upon the words of the gold seekers and from his own considerable knowledge, he walks us through the streets of this once booming, jumping town, as if we were there on any old Sunday, the wildest day of the week, strolling by saloons and fandango halls, our ears ringing with a multitude of sounds. From there, we can imagine entering into the world of the Gold Rush as seen through the eyes of those who lived it. Further on in his narrative, he gives us a tour of the venerable Mariposa County Courthouse, built in 1854 and still standing today. This two-story structure served as the courtroom for one of the most famous legal cases in nineteenth-century California history, *Biddle Boggs v. Merced Mining Company*.

Many more vignettes of California life during this rambunctious era are presented by our Sierra raconteur in this beautifully documented and illustrated book. Gary Noy and Heyday are to be congratulated for producing this lively and engaging account of a pivotal event and time that made California a worldwide symbol of golden opportunity. Now, dear reader, enjoy these "49 Tales of Seekers, Scoundrels, Loss, and Luck."

Gary F. Kurutz is the author of The California Gold Rush: A Descriptive Bibliography of Books and Pamphlets Covering the Years 1848–1853 *and has been the Director of Special Collections at the California State Library since 1980. He has also held positions as Head Librarian at the Sutro Library, Library Director at the California Historical Society, and Bibliographer of Western Americana at the Huntington Library. Kurutz has lectured widely on the Gold Rush. He is the executive director of the California State Library Foundation and a curator emeritus at the California State Library.*

PREFACE

It is a place of life, death, mystery, and windswept remembrance.

Hornitos, in the southern region of the Mother Lode near Mariposa, was founded in 1850 by Mexican miners banned from the nearby town of Quartzburg. Their little camp was named for the unique Mexican mounded burial tombs found in its cemetery: "Hornitos" translates to "little ovens," which the graves were said to resemble. It is surrounded by gently sloping hills that showcase undulating chocolate-brown stone walls enclosing checkerboard fields. Slate outcroppings stand like primordial gravestones. Once, the town of the little ovens was one of the richest and toughest of the Gold Rush mining camps, with fifteen thousand eager and confident inhabitants. An estimated $40,000 in gold departed the boomtown daily. In its heyday, Hornitos supported four hotels, six fraternal lodges, half a dozen general merchandise establishments, and innumerable saloons and fandango halls. As the mines played out, the residents drifted away, leaving a semi–ghost town. Today most of Hornitos is abandoned and only a tiny permanent population of less than a hundred endures. The streets are only a few feet wide, with barely enough room for a car to squeeze through. Shuttered, derelict structures spot the grassy knolls. Many of the town buildings are historic and in a state of elegant arrested decay.

A potholed path climbs to the starchy white St. Catherine's Catholic Church, crowning a rise. The church was built circa 1865 and is buttressed with stone. The exposed heights offer a commanding view of Hornitos and the surrounding landscape but also a unique portal into the past, for here, in every direction and within easy view, and more than anywhere else in Gold Country, are tantalizing glimpses of all the themes that animated the cultural eruption known as the California Gold Rush. Here is a panorama of hope, disillusionment, discrimination, grit, reinvention, violence, and dissipation. Look west and you can imagine optimistic wayfarers on the winding ribbon road leading to a land of golden fantasy, as well as disappointed argonauts exiting by the same dusty route. Gaze into the village nestled below and you can almost hear yesterday's rattling whiskey glasses, the creak of the stagecoach, the faint echoes of guitars, a symphony of many languages and dialects, the laughter of the fortunate, and the sobs of the downtrodden. Here your mind's eye can visualize the rumored subterranean passageways used as a hideout for the legendary outlaw Joaquín Murieta. Here, in this sleepy hamlet, there was blood in the streets, fear of the unknown, profit in the stores, and hymns and curses floating on air. A cemetery extends behind the sanctuary, and there we find weathered wooden slabs, lovingly hand-scratched with the names of the departed, as well as more-elaborate marble monuments carved with remembrances of those who came seeking their fortunes from every corner of the world: Mexico, Chile, Germany, Switzerland, Italy, Canada, Austria, Ireland, France, and Denmark. There are markers for country boys from Maine, city slickers from Pennsylvania, hardscrabble farmers and innocent romantics from throughout the States. Here, within the shadow of the peaceful white church on the hill, one looks upon the final resting place of those who found adventure and a home, and, for some, the remains of a dream that died. This one little town offers a rich Gold Rush tapestry, a precious album of historical variety.

As the town of Hornitos symbolizes the wealth of stories emanating from the California Gold Rush—stories under our feet, stories tantalizingly close but distant in time and attitude—the key word in all this is "stories." This book, *Gold Rush Stories*, is not (and is not designed to be) a traditional, linear, chronological history of the Gold Rush. Instead, it is an attempt to corral the amorphous blob that embodied the social, economic, political, and personal aspects of this seminal event by focusing on the tales and accounts of the participants. In letters, diaries, journals, letter sheets, and newspaper articles, the Gold Rush participants expressed their aspirations, discoveries, frustrations, losses, homesickness, determination, and despair. Often these stories were flavored by the prejudices of their tellers, and we should remember that they should not necessarily be considered as representative of a larger group. The memories recorded here are presented unembroidered, complete with misspellings, obscenities, vulgarity, and bigotry. Any other approach would be dishonest. Just as many nationalities seasoned the spicy cultural casserole of the Gold Rush, many divergent voices added to the rousing but cacophonous chorus. This heady mixture reinforces the California Gold Rush as not simply a distinctive expedition in search of hidden treasure but also an experiment, an improvisation, and a dazzling, perplexing social kaleidoscope that is hard to classify.

The time period into which we should confine these stories is one of the ongoing and unresolved challenges of the California Gold Rush. Determining when the Gold Rush began is simple; deciding when it ended is difficult. The Gold Rush was detonated by the discovery of shiny flakes and nuggets in the tailrace of Sutter's Mill at Coloma in January 1848, but historians disagree on its finale. It can be persuasively argued that the Gold Rush has never ended, as the influence of entrepreneurial risk-taking that characterized the episode continues as a venerated California trait. Some have attempted to pin down the Gold Rush's conclusion as the moment when more gold seekers were leaving California than entering—roughly 1855—but this approach dismisses the associated endeavors that continued unabated thereafter. There is no simple resolution and, for the purposes of *Gold Rush Stories*, we are studying a Gold Rush "generation," focusing as much as possible on events from 1848 to 1865, while keeping in mind that this bracket of dates is necessarily fluid; many of the stories here have postscripts, some of which resonate into the present day.

What to call the geography of the California Gold Rush is tricky as well. The gold-bearing region has several names: some simply call it "Gold Country," while participants in the Gold Rush frequently referred to the area as "the goldfields" or "the diggings" or "the placers."

Although the term "Mother Lode" was originally applied to a limited area—about 120 miles from Bear Valley to Auburn—many sources (including this book) use it to refer to the entire region, just as "Veta Madre" was used for centuries to describe rich veins of gold or silver anywhere in the Spanish-speaking universe.

The placer deposits in the Mother Lode are rich and of particularly high quality, and in the earliest days of the Gold Rush the yield of surface riches was extraordinary—estimated at $81 million in 1852 alone. After the placer gold was quickly exhausted, vastly more productive underground mining took precedence. Statistics from the California Department of Conservation's Divisions of Mines and Geology indicate that from 1848 to 1967 California was the source of more than 106 million troy ounces of gold—enough gold to form a cube twenty feet high, twenty-five feet wide, and twenty-five feet deep.

But the landscape was much more than a natural-resource treasure house. Gold Country was also the realm of uncontrollable fantasy, a province promising freedom and opportunity, and an escape from the humdrum. And yet the California Gold Rush offered a dynamic quest that was always shadowed by the specter of failure and disaster. It was a monumental gamble, what philosopher Henry David Thoreau called "the world's raffle." But, for most, the greatest failure was not in risking the journey and going bust in a faraway land, but in not taking a chance in the first place.

In the pages that follow, stories of the California Gold Rush await. Let's begin our journey in Monterey, California, as two men board a ship bound for Peru. It is August 1848 and, in their hands, they carry the gleam of the future.

THE WINGS OF THE FUTURE
BEGINNINGS

On August 30, 1848, two men boarded a sleek three-masted sailing ship in the sparkling harbor of Monterey, California. Each carried a parcel whose contents would change the course of history. These were not ordinary packages but vessels of wonderment, miraculous boxes of magical enchantments that possessed the extraordinary ability to veil thoughts, cloud judgment, and move mountains. This unique cargo contained the catalyst of hope, the glimmer of optimism, and the harbinger of wild imaginings that would usher in the largest human migration to a specific location and for a single purpose in more than six centuries. These parcels contained gold.

Lieutenant Lucien Loeser, a West Point graduate from Pennsylvania, carried an astonishing letter and a receptacle variously described as an "oyster tin" or a "tea caddy" containing 228 ounces of gold flakes, spangles, and nuggets. David Carter, a businessman, had a chest bearing 1804.59 troy ounces, or just over 123 pounds, of gold dust.

From Monterey, the duo sailed on the *Lambayecana* to Payta, Peru, boarded a British coastal steamer to Panama, scrambled across the isthmus, secured passage to Jamaica, and finally pushed on to New Orleans, arriving November 23. David Carter was interviewed by the *New Orleans Daily Picayune*, but the reporters were disappointed that the businessman did not fan the flames of curiosity that were sweeping the continent. Carter assured them that finding the gold was hard work, miners had died in the pursuit, and the amount of gold in California was exaggerated. Not content with Carter's description, the *Picayune* editors added a postscript from the August 14 edition of the *Monterey Californian* contradicting Carter: gold was readily available, ripe for the picking, not hard work at all.

The excitement was not contained in New Orleans. News of the abundance of gold was already being delivered eastward by U.S. Navy Lieutenant Edward Fitzgerald Beale, and rumors of gold strikes had been floating for weeks as far west as Hawaii, down under in Australia and New Zealand, and in Chile, Mexico, and Peru. But these were merely speculations. Loeser and Carter's parcels would provide hard proof.

David Carter delivered his chest of gold to the United States Mint in Philadelphia on December 8, 1848. Lucien Loeser deposited his container at the mint one day later. The gold was assayed, and a few days afterward the director of the mint, Robert M. Patterson, reported to the secretary of the treasury, Robert Walker, that the gold possessed an above-average fineness rating of 894—"slightly below the standard fineness, which is 900." The 900 ranking was the benchmark used for most gold bullion coins of the era. Patterson added that the value of the two deposits combined was $36,432. (In today's dollars that would be $1.1 million.)

Along with his parcel of gold, Lt. Loeser also carried with him a letter from California's military governor, Colonel Richard Mason, written with assistance from his adjutant, William Tecumseh Sherman, to General Roger Jones, the adjutant general in Washington, D.C. The four-thousand-word report detailed Mason's visit to the goldfields two hundred miles east of Monterey at the foot of the massive Sierra Nevada range. Mason told of countless watercourses and easily accessible gulches "which contain more or less gold." He continued, "Those that have been worked are barely scratched, and, although thousands of ounces have been carried away, I do not consider that a serious impression has been made upon the whole. Every day was developing new and rich deposits." Colonel Mason argued that "rents or fees" could be collected from those wishing to mine these lands controlled by the government, but, after careful consideration, he chose a different

Colonel Richard Barnes Mason: California's military governor when the Gold Rush began. Portrait by Bradley and Rulofson, c. 1850. Courtesy of the California State Library, Sacramento; California History Section.

course: "I resolved not to interfere but to permit all to work freely." Filled with evocative detail and eyewitness accounts, the letter featured impressively precise reckonings of gold both mined and anticipated. The report was so persuasive that President James K. Polk did not wait for the assay results to validate the golden plenty of California to an eager public. On December 5, 1848, based upon Colonel Mason's account, Polk proclaimed in his annual State of the Union address:

> It was known that mines of the precious metals existed to a considerable extent in California at the time of its acquisition [into the union in 1848]. Recent discoveries render it probable that these mines are more extensive and valuable than was anticipated. The accounts of the abundance of gold in that territory are of such an extraordinary character as would scarcely command belief were they not corroborated by the authentic reports of officers in the public service who have visited the mineral district and derived the facts which they detail from personal observation.

The moment had arrived. Society had been exposed to a formidable contagion: gold fever.

This powerful force verged on fanaticism and appeared unshakeable. Nineteenth-century historian Hubert Howe Bancroft fervently exclaimed that gold fever "touched the cerebral nerve that quickened humanity, and sent a thrill throughout the system. It tingled in the ear and at the finger-ends; it buzzed about the brain and tickled in the stomach; it warmed the blood and swelled the heart." Twentieth-century scholars T. H. Watkins and R. R. Olmsted felt that, "like some kind of hallucinogen, gold heightened, distorted, and accelerated reality, propelling spurts, bursts, and great explosive movements. The chronicle of the years that accompanied and followed the Gold Rush is not so much a history as the record of a seismograph."

Gold was the attraction that fed the fever, but what gold symbolized provided the motivation for a cohort of yearners to leave their homes and seek the end of the rainbow. By foot, wagon, and ship, they would journey beyond yonder valley, climb the rugged mountains, cross the seemingly endless American outback, hike the steaming jungle paths, and cross the swelling seas to a mysterious land far away, where their golden reward, plentiful and free, waited patiently for their arrival.

The world as they knew it was about to change. An extraordinary adventure beckoned that followed uncharted and peculiar paths. Most expected golden glory but found heartbreak instead.

UNDISCIPLINED SQUADS OF EMOTION

MOTIVATIONS

In a July 19, 1850, letter to his mother, a twenty-nine-year-old gold seeker from New York named William Swain attempted to characterize the motivations of those who sought gold in California. He identified four classes of searchers. There were the risk-takers, for whom, wrote Swain, "mining has all the excitement of gambling"—men who excitedly viewed the hunt as they might "buy a lottery ticket in the hope of drawing the highest prize." Second were those escaping poor-paying jobs with even poorer prospects, those who were "not disappointed at being their own masters." Third were the rogues seeking unbridled freedom in a life in which, "without fear of punishment, they can drink, fight, and gamble, and, indeed do anything except steal and murder." And, finally, Swain concluded, there were "a better class of men" who had nothing to lose, since "no better lot opens before them."

Swain's list is a good starting point, but the stimuli were far more numerous. Often the motivations were complex, complicated, and deeply rooted in both personal strivings and perceived familial and societal responsibilities. But sometimes the rationales were humble, even mundane. Underlying individual inspiration was a powerful thread that connected the national psyche, and the American West represented unrestrained freedom to pursue those dreams.

Frequently, a man's intention—and they were almost always men—was simply to improve his family's lot in life. A short stint in the remote cradle of prosperity would mean more money, more land, less hardship, and a happier existence. On occasion, this goal was supplemented by the desire to break the bonds of seemingly intractable social strata and prove the neighbors wrong. Nathan Chase expressed as much in a letter to his wife in March 1852: "Jane I left you and them boys for no other reason than this to come here and procure a littl property by the swet of my brow so that we could have a place of our own that I mite not be a dog for other people any longer." Ohioan David DeWolfe echoed that sentiment: "[I hope] to make enough to get us a home & so I can be independent of some of the darned sonabitches that felt themselves above me because I was poor cuss them I say. . . . Darn their stinking hides. If God spares my life I will show them to be false prophits."

The motive for some was an escape from debt. In 1852, James Pierpont composed a song entitled "The Returned Californian" that addressed that theme. One verse read:

> Oh, I'm going far away from my Creditors just now,
> I ain't the tin to pay 'em and they're kicking up a row.
> I ain't one of those lucky ones that works for 'Uncle Sam,'
> There's no chance for speculation and the mines ain't worth
> a [damn].

For many, the drive was fueled purely by money. Aside from those who went to the mines and the nearby boomtowns themselves, other, less-adventurous souls, or those who lacked the opportunity, the gumption, or the funds to set out for California, hoped to profit nonetheless by furnishing financial assistance to "substitutes" who agreed to share their proceeds upon return from the goldfields. In the canyons of Wall Street, the Gold Rush was good business. As the *New York Herald* reported on December 10, 1848, "The California gold fever rages in Wall street to an enormous extent, and fancy stocks have taken an upward start upon the strength of it."

California miner Charles Nash, originally from Michigan, wrote to his hometown newspaper in 1850 that he believed the tales of the abundance of gold were a sham: "Such stories are started by merchants and dealers in liquors and provisions, etc. to attract a crowd to their place and make a market for their goods." For those already in residence in California, however, the opportunity for financial reward seemed endless. Sam Brannan would become the wealthiest man in California, not by mining but by charging exorbitant prices for consumer goods that miners found necessary and desirable in San Francisco, Sacramento, and various goldfield settlements. Brannan constructed flour-mills, opened a hotel, built warehouses, published San Francisco's first newspaper, and purchased empty lots and abandoned buildings that he later sold or rented for ridiculous gains. He also launched farming centers to help feed the hungry multitude. Usually the earliest operating establishments in a district, Brannan's enterprises were regularly the first venues offering blankets, boots, gold pans, knives, tea, coffee, liquor, and many other commodities to newly arrived gold seekers—all at a hefty profit, of course. Sam's wife, Ann Eliza, efficiently summarized his inspiration in an 1848 letter to her friend Mary. As she put it, "Now is the time to make money."

A few argonauts were primarily motivated by the hope of social and racial justice. In 1849, Vicente Pérez Rosales, a gold seeker from Chile, referred to the transpiring events as "the golden dream." California represented a new dawn, free of social rancor and open to all. The forty-niners, Rosales noted, "had arrived from all parts of the world, flocking to the great fair nature had opened up for the human race. They had come to a land where . . . the generous immigration laws had removed the word stranger from the vocabulary." For African American slaves and freedmen, California represented a prospect unavailable elsewhere in the United States. In the goldfields, there was a greater opportunity to gain freedom or earn enough to purchase the freedom of their enslaved relatives back East. Their numbers were small. The census of 1850 counted only about one hundred black people in the state, but the population increased steadily, to around twenty-five hundred—including ninety women—in 1852, many of them drawn by the chance for personal liberty. John Elza Armstrong, a white forty-niner from Ohio, recorded in his diary, "I saw a colored man going to the land of gold prompted by the hope of redeeming his wife and seven children. . . . Success to him. His name is James Taylor." Upon arrival

To St Louis ... 350 Miles
To California 1700 Ms
THE INDEPENDENT GOLD HUNTER ON HIS WAY TO CALIFORNIA.
I NEITHER BORROW NOR LEND

Seized by gold fever, forty-niners headed west carrying their gear and their dreams, as illustrated in this whimsical 1850 lithograph, "The Independent Gold-Hunter on His Way to California." Kelloggs & Comstock, English & Thayer, 1850. Courtesy of the California State Library, Sacramento; California History Section.

in the mining camp of Ophir, near Oroville in Butte County, Armstrong observed other African Americans, including a miner from Texas who hoped to eventually buy freedom for his wife and children and relocate to New York or Massachusetts.

A handful of emigrants were inspired by religious fervor. Pouring uncontrollably into "heathen" California—as it was regarded by many spiritual leaders in the urban and industrial centers of the Atlantic seaboard—were legions of the unsaved, the dissipated, and the lost. It was the hour, they argued, to take up the weapons of righteousness and assemble a new crusade to rescue the gold seekers from themselves. In 1853, William Newell, a Presbyterian minister from Liverpool, New York, gave these marching orders to members of his congregation headed to California:

> Arise then, my brethren! Go forth toward the setting sun, and there be sure you *shine*.
>
> Let the light of your Christian life spread through the valleys of the Sacramento and San Joaquin; let it gild Sierra Nevada's tops. . . . Leave not your stamp in the gold and granite, the hills and valleys of California that crumble away, but leave it in something vastly more valuable and lasting. . . . Go, then, dearly beloved of the Lord! May it be yours to wave the banner of the cross over the El Dorado of our world.

For the less hardy, it was the weather that lured them. Rumors of the goldfields as a tropical paradise were appealing to those from frozen climes. In 1849, Dr. Felix Wierzbicki, a Polish-born medical doctor and Mexican War veteran who published *California As It Is, and As It May Be*, prescribed "as a medicine, to all vinegar-faced, care-corroded gentry that are well to do in the world, to come and settle in the rich valleys of California, where good health and azure skies can be enjoyed; where winter does not touch you with its freezing hand."

For the overwhelmingly young and male participants in the Gold Rush, the spur to head west could be as modest as impressing a waiting sweetheart or irritating their parents in a burst of rebellion or even wishing to surpass the accomplishments of a sibling.

But perhaps the motivating factor that drove the greatest number of emigrants during this movement was best expressed in two simple words written by Sarah Royce. Sarah was the mother of Professor Josiah Royce, a prominent nineteenth-century Harvard philosopher and social commentator. Josiah's parents had traveled from Iowa to California in the exhilarating days of 1849, arriving after a difficult six-month passage by prairie schooner on the inhospitable road to the land of golden dreams. As she recalled in her memoir *A Frontier Lady*, Sarah was disheartened as she crossed the burning Forty Mile Desert. Her young daughter, Mary, was sick and there had recently been a cholera outbreak in the wagon train. Water supplies had all but evaporated as Mary begged, "Mamma I want a drink." There was "but a few swallows left," and Sarah silently repeated over and over, "Let me not see the death of the child."

At that moment an odd incident occurred, an event Sarah interpreted as a divine sign. In the heat of the noonday sun, with no other fires visible, Sarah recalled that "just before me to my right a bright flame sprang up at the foot of a small bush." The sagebrush blazed but a few seconds and "then went out, leaving nothing but a few ashes and a little smouldering trunk." To Sarah, it was an omen, a heavenly beacon of hope. Strengthened by the flaming prophecy, Sarah and her husband resolved to continue westward, Mary regained her health, and a few years later, the couple gave birth to their youngest son, Josiah Jr., in the rough mining town of Grass Valley in Nevada County.

Prior to their arrival there in 1854, the Royces had barely subsisted in several other Gold Rush camps; Grass Valley was their ninth home in five years. Despite the struggles they encountered, the Royce family never dismissed Sarah's vision of a brighter future waiting in California. Her two-word declaration was simple, and those words could serve as the mantra for all who came to the land of golden possibilities. They made the journey, Sarah said, because California was "something better."

THE CONSEQUENCE OF FEARFUL BLINDNESS

TWILIGHT OF THE CALIFORNIOS

Reflecting on the decades prior to the Gold Rush, gauzy-eyed romantics often claim that the charmed world of the rancho defined the Mexican province of Alta California. In their eyes, ranchos were gracious societies of endless fiestas, elaborate weddings, colorful quinceañeras, and rousing fandangos. The women were stunning, delicate, and sophisticated, and the men were robust, noble, and extraordinarily skillful and daring horsemen. The rancho landholdings were massive and open, stretching across tens of thousands of green rolling hills, and roamed by huge herds of cattle, sheep, and horses. All were happy and prosperous, living in a grand style reminiscent of the wealthy Spanish hidalgos. It was a privileged culture of cordial hospitality, selfless generosity, and elegant gentility. While this vision had elements of truth, mostly it was a fantasy, fashioned from smoke and mirrors. In reality, this fairytale was built on a shaky foundation of brutal treatment of the Native people, extensive debt, clashing allegiances, and political uncertainty. This illusion would be irrevocably transformed by the cataclysm of the Gold Rush: whether chipped away acre by acre or swept up in massive tracts, most of the land held by Californios had new owners by the end of the nineteenth century.

The Gold Rush not only sparked monumental social upheaval but significant legal transformation as well. Californios—residents born in

California to Spanish or Spanish-descendant parents during the period of Mexican control—frequently saw their rights and influence swept away by the insatiable hunger for land of the forty-niners. California's political transition from Mexican authority to American rule following the Mexican War of 1846–48 fundamentally altered legal traditions and procedures in California, changing the way land was surveyed and titles were held. Californios and holders of Mexican land grants would reap a terrible harvest from these changes.

Before the Mexican War, hand-drawn maps called diseños served as land title documents. "Diseño" translates as "design," and these often beautiful maps marked imprecise property lines according to geographical attractions within the area. A typical diseño might include drawings of a massive oak tree, a streambed, or a rock outcropping. An old joke typified the style of the diseños: "Our property extended from the rock that looks like a bear to the bear that looks like a rock." More exact and detailed maps were considered unnecessary in the largely unpopulated region, and if there were boundary disputes, they could be resolved by the application of patience, additional description and testimony, and some neighborly negotiation. While some land grants were held by non-Californio residents, such as John Sutter, a German-born Swiss immigrant, the vast majority of the disputed land was controlled by Californios—a total of fourteen million acres in the years before the Gold Rush.

But as the swarm of new settlers swept in, they frequently disregarded Californio property rights and claimed the land for themselves as squatters. With the territory now under a new legal system, the official township and section survey maps replaced the diseños as the legitimate title documents. The 1848 Treaty of Guadalupe Hidalgo, which ended the Mexican War, had promised Californios "the free enjoyment of their liberty and property," but this guarantee was promptly ignored, and clashes over competing land titles were inevitable.

In 1851, California senator William Gwin introduced what was commonly called the Land Law of 1851, designed to find equitable solutions in an atmosphere of vagueness and uncertainty in the aftermath of the Treaty of Guadalupe Hidalgo. Gwin vigorously touted its benign nature in speeches to Congress in late 1850 and early 1851. The proposed act, Gwin noted, would save the Californios "from alarm, agitation, and despair" by setting up a three-member commission to consider evidence presented by any titleholders who appeared before it. The act passed easily on March 3, 1851. The immediate consequence was suspicion, turmoil, and decisions deferred.

With fourteen million acres at stake, Californios and other land-grant holders were forced to prove their ownership in court with the outmoded, invalid diseños. The most contemptible players in this drama were the cunning lawyers and swashbuckling speculators—men like Horace Carpentier—who came to California during the rush and hoodwinked trusting Californios whilst posing as allies and angels. Resolution of these cases averaged seventeen years, and they racked up massive legal fees. Some landowners were forced to sell or mortgage significant portions of their estates to meet the rapidly mounting expenses. While some Californio domains remained in family hands, an estimated 25 percent of Californios lost all their land within the first decade of the Land Law, either through rejected claims or to pay off outrageous legal fees. Additional lands were lost in subsequent years due to continued legal challenges or inadequate income to meet rising taxes and high mortgage payments and maintenance costs.

Born in New York in 1824, Horace Carpentier set out for California in 1849, one year after graduating from Columbia Law School. Horace was not interested in gold; he thirsted for land. Based in San Francisco, Carpentier spent two years practicing law throughout Alta California, and in 1851 his unfulfilled desire for land led him to covet property hugging the eastern edge of San Francisco Bay. Much of the East Bay was made up of the nineteen-thousand-acre Rancho San Antonio, owned by the Peralta family. Carpentier and his equally devious partners persuaded family patriarch Vicente Peralta to trust Horace as their friend and agent. Carpentier curried favor by claiming piety and fidelity to the family's values and interests; he prominently wore a cross around his neck, conversed endlessly about his faith, and fervidly described his program for providing the Peralta estate with security and increased wealth. It was a con.

Vicente Peralta, like other Californio landowners, was stunned by the rapid social and legal changes accompanying the Gold Rush influx of Anglo Americans. He wisely realized the need for a more knowledgeable figure to manage his affairs and protect his title. Unfortunately, he trusted Horace Carpentier. Behind his back, Carpentier crafted a lease agreement with squatters and speculators that required a mortgage against the Peralta family to pay operating expenses and legal fees, with the spoils benefitting Carpentier himself. This arrangement could strip the Peraltas of ultimate control of their lands, and when presented to Vicente Peralta for his signature, he refused to sign. In this period of roiling land-title disputes, this plunged Peralta property—parts of which were already being coveted and contested by others—into legal limbo, which allowed Carpentier to swoop in and purchase a large swath of the East Bay waterfront for pennies.

On May 17, 1852, Horace Carpentier was granted control of the entire waterfront for a period of thirty-seven years (later altered to "in fee simple forever") in exchange for payment of $5 and a promise to build three wharves and a schoolhouse. The Peralta family lost their land and received nothing in return. Carpentier established a ferry monopoly and operated a toll bridge over an estuary on the shoreline. Virtually every person, animal, commercial item, and mail parcel using the bridge required a fee payable to Carpentier and his henchmen.

At Carpentier's insistence, the portion of Rancho San Antonio then known as Contra Costa, including the waterfront he now controlled, was rechristened as Oakland. Carpentier was elected to the California State Assembly in 1852, and he immediately began lobbying to incorporate Oakland as a city. Two years later, twenty-nine-year-old Horace Carpentier was elected the first mayor of Oakland. Both elections were widely regarded as colossally fraudulent. Carpentier entrenched his supporters in positions of political and civic authority, Oakland became his personal playground, and he became rich and powerful.

Although he failed in a bid to become the Democratic nominee for the state attorney general, Carpentier's political influence continued unabated in Oakland for years. He became president of the California State Telegraph Company, the first to provide telegraphy to the state, and he was later president of the Overland Telegraph Company, which provided the initial telegraphic link from California to the East Coast. Carpentier was also a founder and director of the Bank of California. In October 1877, the *Oakland Daily Transcript* provided this assessment of Carpentier's career: "If the early settlers had taken Horace Carpentier to a convenient tree and hung him, as they frequently threatened to do, the act would have been inestimably beneficial to immediate posterity."

Upon his eventual demise in 1918 in New York, Carpentier bequeathed millions of dollars to Columbia University and Barnard College and an additional several hundreds of thousands to the University of California and the Pacific Theological Seminary in Berkeley. He left nothing to Oakland.

The Peralta family, who unsuccessfully pursued their claim on their land until 1910, received only a single tribute: a street in Berkeley, Peralta Avenue, was named for them.

Horace Carpentier's audacious swindle of the Peralta land is the sort of dramatic tale of fortunes won and lost that one might expect from a Gold Rush story. But even more tragic, and more emblematic of the piece-by-piece loss that most Californios experienced, is the story of Mariano Guadalupe Vallejo. In 1890 he was a little old man who spent his days puttering around his prefabricated Carpenter Gothic–style house in Sonoma, doing chores, tinkering with equipment, and tending a handful of grapevines on his rancho. In the preceding decades he had read voluminously from his library of thousands of books and had written and rewritten a five-volume history of California. This unimposing, stooped figure with thinning hair and bushy gray muttonchops was once the most powerful land baron in Northern California.

At the height of his influence, Mariano Vallejo controlled 175,000 acres and was the trusted advisor of Mexican governors and military leaders. Prior to the Gold Rush, Vallejo was Commander of the San Francisco Presidio, Comandante of the Fourth Military District, and Director of the Colonization of the Northern Frontier, the highest-ranking military positions in Alta California. He had the sole authority to issue land grants in the region, he commanded the security arrangements, and he dominated the cultural landscape. He founded the town of Sonoma, his center of operations; established the growing community of Vallejo; and named Benicia, a nearby port city that briefly served as California's state capitol, after his wife, Francisca Benicia Carillo de Vallejo.

His fortunes changed when the Norte Americanos claimed ownership of California. Vallejo was arrested in the 1846 Bear Flag Revolt and withered to ninety-six pounds in jail. Following his release during the Mexican War, he announced his undying allegiance to the new United States authorities by ceremonially burning his old Mexican uniform. Vallejo was a participant in the California Constitutional Convention of 1849, was elected to the first California State Senate in 1850, and offered the town of Vallejo as the site of a permanent state capitol.

Before the Gold Rush, his extensive estate harbored hundreds of employees, ten thousand cattle, six thousand horses, and thousands upon thousands of sheep. He maintained several elaborate residences and had far-reaching and profitable commercial interests, such as leather tanning, that fed a lavish lifestyle that was occasionally propped up by mortgages on the landholdings. It was all utterly destroyed, however, when the onslaught of fervid gold seekers reached a fever pitch. Vallejo's lands were overrun by argonauts and squatters, and his livelihood was devastated. Livestock were stolen and slaughtered by the thousands; trespassers came and milked his cows at night. Vallejo's property titles were questioned and his business enterprises were swept away by the tidal wave of the Gold Rush. Endless proceedings to verify his land ownership ensued, and Vallejo's remaining holdings and fortune quickly disappeared. His 175,000-acre domain shrunk to a few hundred acres. What was left was mismanaged by his son-in-law John Frisbie, who lost the last of Mariano's property in the late 1860s, although he managed to regain ownership of Vallejo's small Sonoma homestead in 1871. But by then Mariano Vallejo was careworn and resigned to his fate, lingering in the twilight of history. Occasionally, it is believed, Vallejo would stroll a few blocks to Sonoma Plaza, his former headquarters, and meander through the grounds, recalling days of yore. "If the Californians could all gather together to breathe a lament," he waxed nostalgically in a January 11, 1877, letter to his son Platón,

Mariano Guadalupe Vallejo was among the most prominent and powerful of the Californios. Courtesy of the California State Library, Sacramento; California History Section.

"it would reach Heaven as a moving sigh which would cause fear and consternation to the Universe. . . . This country was the true Eden, the land of promise."

Mariano Guadalupe Vallejo's Sonoma estate was named for the several creeks that flowed on his spread. The Native name was Chiucuyem, or "the crying mountain." Vallejo called it Lachryma Montis—"the tears of the mountain." It is likely that Vallejo, a proud and dignified man, shed many tears as he remembered the long-ago, fading memory of the "true Eden." Mariano Guadalupe Vallejo died in 1890. He was eighty-two years old. Vallejo's obituary was front-page news and was generally treated as a paean to a lost culture: the fabled Spanish California, now an illusory, phantom society.

"I KNOW IT TO BE NOTHING ELSE"

THE SAD TALE OF JAMES MARSHALL

One of the men who had the most to gain from the Gold Rush died embittered and broke. His chance discovery on a frosty morning in January 1848 transformed his life, revolutionized Gold Country, and altered the course of world history. But, for him, the moment brought only endless heartache. It was, he wrote near the end of his life, "a discovery that hasn't . . . been of much benefit to me." His name was James Wilson Marshall.

Marshall, a transplant from New Jersey by way of Missouri, arrived at Sutter's Fort in July 1845 and was quickly hired by John Sutter, the imperious master of a settlement called New Helvetia. A clever jack-of-all-trades, Marshall could be diffident and sullen. He was a difficult fellow to like, but he quickly became Sutter's favorite for carpentry and construction. Sutter described Marshall as "a very curious man, [who] quarreled with nearly everybody though I got along with him very well. He was a spiritualist. He dressed in buckskin and wore a serape. I thought him half crazy."

As New Helvetia grew, Sutter's kingdom needed an inexpensive and convenient supply of lumber. He asked Marshall to find an appropriate site for a sawmill in the Sierra foothills nearby. After considering several locations, James Marshall chose the Coloma Valley, straddling the American River about thirty miles east of Sacramento. The valley nestled on the cusp of the Sierra Nevada, with abundant forest reserves nearby. It seemed perfect, and Sutter was pleased.

John Sutter and James Marshall entered into a partnership—Sutter provided the capital, and Marshall would construct and manage the mill. A land-lease agreement was reached with the local Yalesummi tribe, and Marshall and his team of mostly Mormon laborers and artisans began construction of the sawmill in 1847.

The mill was finished by the end of the year, but a problem arose. The tailrace—that is, the drainage ditch leading away from the water-powered mill—was too shallow. Water could not efficiently flow through, leaving the millwheel unable to function correctly. In order to make the giant wooden paddlewheel turn properly, the tailrace had to be deepened.

James Marshall's workforce pried boulders free and shoveled away dirt. At night the river was rerouted through the ditch to loosen the soil and wash away debris. Then, while inspecting the previous night's progress one chilly morning in mid-January, Marshall found a pea-sized gold flake glimmering in the tailrace. And then he saw another and another as morning light spread across the mill.

Marshall would later write of that moment, "I DISCOVERED THE GOLD. I was entirely alone at the time. I picked up one or two pieces and examined them attentively; and having some general knowledge of minerals, I could not call to mind more than two which in any way resembled this—*sulphuret of iron,* very bright and brittle; and *gold*, bright, yet malleable." "It made my heart thump," he recalled, "for I was certain it was gold." Marshall sat down to "think right hard" about how to proceed.

> I . . . went up to Mr. Scott (who was working at the carpenters bench making the mill wheel) with the pieces in my hand and said, "I have found it."
>
> "What is it?" inquired Scott.
>
> "Gold," I answered.
>
> "Oh! no," returned Scott, "that can't be."
>
> I replied positively,—"I know it to be nothing else."

Years later, Marshall remembered the date as January 19, 1848, but the more detailed diary of employee Henry Bigler set the day as January 24, which has become the historically accepted day of discovery.

Marshall hurried to inform John Sutter in Sacramento. They tested the find at Sutter's Fort, and it was indeed gold. Very pure gold. Concerned that a gold strike would result in an exodus of workers, an incomplete sawmill, and a wasted investment, Sutter and Marshall decided to keep their newly found treasure a secret. But it would not remain secret for long. Samuel Brannan, a merchant at Sutter's Fort, learned of the discovery and carried a small vial of gold dust to San Francisco, proclaiming, "Gold! Gold on the American River!" up and down Market Street, sparking the initial rush. Clever entrepreneur that he was, Brannan waited until he had purchased every available pick, shovel, and pan in the city before making his announcement.

Following the gold discovery, Marshall claimed ownership of the Coloma Valley as a homestead, but his declaration was drowned by a rising tide of passionate gold seekers who did not give a damn about social conventions or legal proprieties. He described the chaotic scene in a letter to the *New York Herald* on June 27, 1849:

> Men soon came to the place where none but a fool or crazy man, they said would go. But alas, they left honesty and honor at home, with a few, very few exceptions. Then commenced a course of rascality, of which Sutter and myself were the principal subjects; at us it was aimed. That many-headed community plundered the persons who had given them wealth by their enterprise! Fourteen yoke of oxen were stolen and butchered and from myself alone, six head of horses, plank and tools were stolen. Indians were set against me who sought my life.

Placer miners seeking gold in its riverbed diverted the American River, and Marshall's mill was left high and dry, and abandoned. In 1853, a government survey reported that the mill had been purchased "with the intention of making walking sticks from its timbers." Marshall turned to prospecting, but he could find no peace after his discovery. He was continually followed by the lazy, the unscrupulous,

James Marshall, whose discovery of gold in 1848 triggered the Gold Rush, died penniless after years of bad luck and questionable decisions. Courtesy of the California State Library, Sacramento; California History Section.

or the neophytes who were counting on Marshall's supposed Midas Touch. In 1908, a "Pioneer" remembered "trailing Marshall":

> Like most gold-seekers, I for one was not satisfied and had the belief that if Marshall (the discoverer of gold) could be shadowed the fountain head would be found. . . . Knowledge having been gained of Marshall starting with a party up between the north and middle forks of the American river, our party of about seven in number followed his, always keeping a sufficient distance in the rear to escape discovery by his party.
>
> It is forgotten how many days his party was trailed, but suddenly one day we came upon him having a siesta about noon. I well remember his appearance and about what he said. . . . He smiled and said that he knew nothing of where gold existed any more than we did. After a short halt our party moved on to hunt further for the precious metal.

Many "trailers" were not as benign as these and threatened violent reprisals if Marshall did not succeed in finding the golden bounty. Fearing for his life, he left Coloma in 1853. In an August 1864 document asking for financial relief, Marshall described the aftermath:

> I was soon forced to . . . leave Coloma. . . . My property was swept from me, and no one would give me employment. I have had to carry my pack of thirty or forty pounds over the mountains, living on China rice alone. If I sought employment, I was refused on the reasoning that I had discovered the goldmines, and should be the one to employ them; they did not wish the man that made the discovery under their control. . . . Thus I wandered for more than four years.

Returning to Coloma in 1857, Marshall, deeply in debt, reluctantly sold a portion of his timber and mill rights to satisfy creditors. He tried prospecting again. No luck. He survived by working odd jobs, and within a few months he had saved enough to buy fifteen acres in Coloma for $15 and started a vineyard. It was successful, initially, and he even earned a medal for his wine at the El Dorado County Fair.

But high taxes and competition led to setbacks. He tried mining once more.

Purchasing a quartz gold mine near the little hamlet of Kelsey, about five miles from Coloma, Marshall and a partner struggled with hard rock mining, and for the most part did not succeed. In the 1860s, Marshall booked a lecture tour to spark interest and investment in the mine, but he was a poor speaker and the tour was a dismal failure. He published a book chronicling his exploits, but the sales were dreadful. James Marshall spent his last pennies on these failed enterprises. He began to drink heavily.

Destitute, angry, and increasingly eccentric in his behavior, Marshall returned to Kelsey and started a blacksmith shop. He petitioned the government for assistance, arguing that as a figure primarily responsible for the wholesale changes in California's economy, he deserved recompense. Others buttressed Marshall's case on humanitarian grounds. In February 1872, the California State Legislature passed "An Act to Appropriate Money for the Relief of James W. Marshall," providing a $200 monthly pension. Renewed several times, the pension ended in 1878 following criticism of Marshall's alcoholism. According to legend, when Marshall appeared before the State Assembly to testify on his own behalf, a whiskey bottle fell from his pocket and crashed to the floor.

With no other income to rely on, Marshall scraped along at his blacksmith shop in Kelsey. He made a few dollars by selling autograph cards featuring his scrawling signature and listing the gold discovery date as January 19, 1848. But he never prospered.

James Marshall died in 1885 at the age of seventy-five. He is buried on a Coloma hillside near his old vineyard and cabin. Philip Bekeart, who as a boy knew Marshall, remarked in 1924:

> All historians have given Marshall credit for being the discoverer of gold in the Sierras, but have handled him roughly in other ways probably because of his mannerisms, his belief in spiritualism, and his drinking. He was so badly treated by the early settlers, who robbed him, and by some of the arrivals of 1848 and 1849 who persecuted him, and he became a

> misanthrope. He was disgusted with the way fate had used him, and with the exception of a few old friends, wanted to be left alone.

Others were not as forgiving. On January 24, 1898, Placerville's *Mountain Democrat* issued a special edition commemorating the fiftieth anniversary of the gold discovery in Coloma. Their characterization of Marshall's final years is scathing:

> [Marshall] scattered his money indiscriminately among his friends and parasites, and he soon became but little more than a common sot. . . . Although naturally worthy of better things, he deteriorated until in plain English, he was unprepossessing in appearance, untidy in person and filthy in habits; his habitation was a den reeking of tobacco juice, strewn and plastered with antiquated quids [plugs of chewing tobacco] and redolent of creosote.

The James Marshall Gold Discovery Monument that marks the spot was the result of a $9,000 legislative appropriation to construct a memorial to Marshall. A bronze-coated zinc statue of James Marshall crowns an ornate pedestal. The figure of Marshall supposedly points to the gold discovery site, but he actually gestures to the right of where gold was originally found.

The inaccuracy is fitting.

In life, gold seekers demanded the impossible from James Marshall: that he magically indicate where to find gold. In death, even his statue continues to fail.

THE TALE OF TELEGUAC

JOSÉ JESÚS, NATIVE RESISTANCE, AND SURVIVAL

In March 1849, a company of newly arrived gold seekers from Oregon raped the women of a Maidu rancheria near Coloma. Incensed, Maidu warriors retaliated, killing five of the attackers on the Middle Fork of the American River at a site thereafter known as Murderer's Bar. Upon hearing the gruesome news, others from Oregon organized an attack on an Indian settlement near Weber Creek, outside of Placerville. A dozen Native men were gunned down and many prisoners were captured.

The captives were paraded to Coloma, where James Marshall, whose gold discovery had triggered the Gold Rush in 1848, was forced by the murderous mob to pick out eight Indians who were best known to him. Some of these men worked for Marshall at the mill in Coloma and belonged to a different tribe from those who had attacked the Oregonians. It did not matter to the vicious horde. Blood would be shed and damn the details.

After a drunken revel, the rabble released their eight captives like rabbits from a cage and ordered them to run for their lives. They were hunted down and seven were killed as they fled. James Marshall attempted to halt the carnage, but as participant John E. Ross recalled, when Marshall "started to advocate the cause of the Indians," angry

Oregonian Niniwon Everman raised his rifle to shoot him. Marshall was given "five minutes to leave the place." James Marshall hastily departed on a borrowed horse.

With the incident concluded, the rabble returned to their diggings. No criminal charges were brought or even contemplated. This was not surprising; a majority of forty-niners viewed the indigenous population as subhuman obstacles to progress, useful only as slave laborers or target practice. Many would echo the sentiments of an acquaintance quoted by diarist William Kelly, an English storekeeper on the American River at the time, who explained, "You know . . . no Christian man is bound to give full value to these infernal red-skins. They are onsoffisticated vagabones; . . . they've got no religion, and tharfore no consciences, so I deals with them accordin."

A few weeks after this mass murder, John Huddart, a Mexican War veteran, shot and seriously wounded a chief of the Stanislaus Siakumne on the streets of Stockton following an alcohol-fueled argument. The founder of Stockton, Charles Maria Weber, arranged for medical treatment for the victim, paying Dr. W. N. Ryer $500 to privately treat the casualty. When the gunshot sufficiently healed, Weber arranged to have the chief gently transported back to his home near Knights Ferry. Huddart, meanwhile, was arrested, tried, and sentenced to three years in prison for the crime. Why was this outcome so startlingly at odds with the circumstances surrounding the slaughtered Indians in Coloma?

The story of the assault on and subjugation of California Natives during the first decade of the California Gold Rush is a well-documented tragedy. Trustworthy, accurate records are difficult to obtain, but deaths due to disease, warfare, extra-judicial execution, and murder numbered in the tens of thousands and perhaps reached one hundred thousand or more; by some estimates, between 60 and 90 percent of the Indian population was destroyed within ten years. What is known for certain is the attitude of most gold seekers and government officials toward the Natives. The policy was extermination, a word that surfaces frequently in discussions and pronouncements from the era. This terrifying attitude not only reflected commonly held beliefs but also primed schemes for further bloodstained campaigns. Ponder these few examples, but a minute register of unnerving thoughts that come up time and again in the historical record:

> It is now that the cry of extermination is raised—a thirst for indiscriminate slaughter rages, and men, women and children, old and young, vicious and well-disposed, of the Indian race, wherever met with, are to be straightway shot down or knocked on the head, their villages plundered and burned and frightened fugitives forced deeper in to the mountains.
>
> *Placer Times* (Sacramento), April 28, 1849

> Idle, thievish, ignorant, degraded and brutish,—thus may we sum up, in short, the character of the California Indian. . . . There will then be *safety* only in a war of extermination, waged with relentless fury far and near. The mining lands must be wrested from the river tribes, and the natives forced to flee into the mountains. . . . Such is the destiny of that miserable race.
>
> *San Francisco Daily Alta California*, May 29, 1850

> The white man, to whom time is money, and who labors hard all day to create the comforts of life, cannot sit up all night to watch his property; and after being robbed a few times, he becomes desperate, and resolves upon a war of extermination. This is the common feeling of our people who have lived upon the Indian frontier. . . . That a war of extermination will continue to be waged between the races until the Indian race becomes extinct must be expected. While we cannot anticipate this result but with painful regret, the inevitable destiny of the race is beyond the power or wisdom of man to avert.
>
> Peter Burnett, "The Governor's Message," January 6, 1851

But this story is not about loss. It is about adaptation and survival among the Native people. It is a tale not of Indians as victims, passively resigned to their fate, but as active participants in determining their destiny in the face of an endless, horrific onslaught. It is a chronicle of resistance, creativity, tenacity, and flexibility. As historian Albert Hurtado so eloquently expressed in his 1988 book *Indian Survival on the California Frontier*, "That so few Indians survived [the Gold Rush] is stark evidence of the prodigious upheavals of a remarkable time. That any Indians survived is testimony that abhorrent conditions can produce courage and strength in people, a tribute to the persistence of humankind." All these compelling threads interweave in the story of José Jesús—chief of the Stanislaus Siakumne, a community within the Northern Yokuts—who was shot point blank on a street in Stockton in 1849.

His birth name is believed to have been Teleguac. A Mission San José record, signed by Father Narciso Durán on November 29, 1834, lists the baptism of a twenty-seven-year-old Tihuechemne of the Northern Yokuts by that name. But he was already better known as Hozá Ha-sóos, leader of the Indian resistance of the "Stanislaus Si-yak-umnas." The Mexican authorities branded him José Jesús, the Chief of the Horse Thief Indians. A little girl who lived in the mission at the time, Luisa Sepulveda Mesa, later recalled that José Jesús was "a handsome young chieftain, who was almost as fair as a Spaniard. He dressed like one. He pretended to be very religious. He knew all the rules of the mission." Charles Maria Weber remembered that José Jesús was a striking figure, six feet tall and habitually dressed in "the full gala attire of the Spanish ranchero, with cotton shirt, and drawers, calzonazos, sash, serape and sombrero."

In the years before his baptism, José Jesús allied himself with two powerful and charismatic Lakisamne Yokuts leaders of Native resistance, Estanislao and Yozcolo. While José Jesús chafed at the missions' brutal domination over Native peoples and their way of life, Estanislao and Yozcolo challenged the episodes of violence and trespass on Native territory that accompanied the increasing Mexican presence in the province.

Estanislao, a former neophyte at Mission San José, was wiry and magnetic. Outraged by the inhumanity he and others suffered within the mission system, Estanislao left San José, established a base of operations in the future Stanislaus County region (which was later named after him), and began recruiting followers. Among those who joined his band were Yozcolo, a rebellious young man from Santa Clara, and José Jesús. Together they forged an effective guerilla strike force, raided Californio livestock herds, persuaded other Indians to abandon mission life, and generally caused havoc for the established order. Three military expeditions were dispatched to neutralize Estanislao. They all failed.

In 1829, a fourth expedition, led by the twenty-one-year-old acting commandant of Monterey, Mariano Guadalupe Vallejo, assaulted Estanislao's San Joaquin Valley fortress of trenches, stockades, and breastworks. After three days, Vallejo's forces captured the fortification, but Estanislao escaped. He sought refuge at Mission San José with Father Durán, who believed that the most effective means to end the Indian resistance would be to pardon its leader. Durán brokered a pardon for Estanislao with Governor José María de Echeandía, and Estanislao remained at the mission until his death from smallpox in 1838. Believing the mission system provided greater safety, many Native Californians and neophytes then returned to the missions. But for some, the situation was never comfortable; resentment festered and resistance resurfaced. José Jesús himself returned to the mission for a time, but it did not end his role as a leader of the resistance.

In 1833, an expedition under the command of Sebastian Peralta was launched to attack Indians in their Central Valley villages. Approaching the foothills east of Stockton, the seventeen members of the Peralta expedition encountered 130 unarmed Mokelumne Indians at a gulch called Sanjón de los Moquelomnes, the "Ravine of the Mokelumnes." Their leader was Chief Cipriano, a comrade of José Jesús. Eyewitness José Francisco Palomares, an expedition soldier, reported:

> We immediately attacked on all sides, firing heavily on them and causing many losses. As they tried to flee at the same time, they made a great crowd at which we fired without fear of miscalculating. Finally, their chiefs seeing that we were overpowering their people, sent to ask for a truce. . . . Peralta ordered us to stop firing instantly. . . .

> Immediately we saw them come out from the arroyo and make a circle, men, women, and children, with arms crossed and eyes lowered in humility. Some of the women carried their children dead or wounded in their arms, and were weeping in a manner to move us to compassion; others scarcely able to stand on their feet came dripping blood.

Commander Peralta ordered his troopers to bind Cipriano and "fifteen of the worst men of the tribe" and tie them to a tree "like a string of beads one after the other." The remaining Indians were freed. Cipriano and the other captives spent the night attempting to free themselves from their bonds. By morning their arms were raw and bloody. Peralta ordered them beaten with "a dozen very severe lashes" as punishment for their attempted escape. The commander lectured the Indians "to remain quietly in the Rancherias . . . if they did not want the white people to exterminate them entirely." Peralta then released them.

This unprovoked and deadly attack upon unarmed villagers incensed José Jesús and he vowed vengeance on the Mexican and Californio hegemony. Never again would his people be humiliated. Ironically, a year after these events unfolded, José Jesús would be baptized by Father Durán, although he remained an active fighter against the mission system.

Yozcolo continued the insurgency as leader and proceeded to bedevil Mexican officials for years, capturing hundreds of horses and raiding mission stores, while José Jesús commenced incursions elsewhere. After a raid at Rancho del Encino Coposo in 1839, Yozcolo was killed and nearly one hundred of his band were surrounded and captured near Los Gatos. The Mexican soldier who killed Yozcolo decapitated him and dragged his severed head into the nearby plaza of Mission Santa Clara, where it was displayed on a pike to intimidate and discourage any additional Indian opposition. It had the opposite effect on José Jesús, however, who rose from the bloodshed to assume leadership of the struggle. Some now called him the chief of San José, and as such he openly displayed contempt for the mission fathers and Mexican authorities, becoming notorious for his defiance and insolence. Luisa Sepulveda Mesa remembered José Jesús catcalling and

Native Californians were immediately and severely impacted by the Gold Rush. Their lives would never be the same. "View of an Indian Rancheria, Yuba City, California," from *Gleason's Pictorial,* March 27, 1852, p. 196. Courtesy of the California State Library, Sacramento; California History Section.

---

sneering at the soldiers in Mission San José during Holy Week, when arrests were forbidden by mission rules.

As José Jesús continued his violent resistance against the missions, the Mexican government, and the Californios, a new group of settlers, the Norte Americanos, were beginning to populate the California interior, even before the discovery of gold. The Americans, as they were collectively identified, did not hate the Californios with the searing hatred of José Jesús, but they distrusted them and saw the Californio ruling class as obstacles to commercial development. While most newcomers also held racist opinions about the region's Native people, a few sought beneficial alliances with them in hopes of protecting their own interests and settlements. José Jesús perceived these potential agreements as a mechanism by which to strengthen his homeland from likely assaults by the hated Californios and Mexican government, and so although he was aware that the Native-controlled lands were shrinking as American influence expanded, he was willing to consider associating with these new arrivals in an effort to carve out a protected domain for his threatened people. Their survival depended upon a tricky balance of adjustment and adaptation.

One of these newcomers was Charles Maria Weber, a German by way of Texas. He was desirous of peacefully expanding his French Camp rancho and bolstering his infant valley settlement then known as Tuleberg, later renamed Stockton. In 1843, Weber met with José Jesús at Sutter's Fort in Sacramento to discuss a possible association that would provide security for both parties and a buffer against unwanted Mexican interference. Weber and José Jesús came to an agreement that lasted for the rest of their lives. They decided that each would establish spheres of influence within defined boundaries and assist each other when necessary. As Frank Gilbert detailed in his 1879 *History of San Joaquin County*:

> The chief devised the building of the American village at the point where it was located, the present site of Stockton, and agreed to provide all the help necessary in the tilling of the soil, and to furnish a war party when called upon to defend the settlers' property against either Indians or Mexicans.

Years before the Bear Flag Revolt of 1846 and the Mexican War of 1846–48, José Jesús linked his people's fortune and future with the Americans in opposition to the old guard, his sworn enemy.

And then the world rushed in. When the California Gold Rush exploded, thousands flocked to the gold-bearing territory of José Jesús, and he and Charles Maria Weber worked together to great advantage in the goldfields. Upward of a thousand Indian miners worked Weber's claims and struck paydirt, which included the discovery of an eighty-ounce kidney-shaped gold nugget that was eventually displayed in the Bank of England. The Weber–José Jesús association is given credit for having established the prosperous Southern Mines of the Mother Lode. In the earliest months of the rush, there were other successful Native miners as well, working independently or as contract laborers. Colonel Richard Mason, in his official government report of August 1848 on mining prospects in the goldfields, estimated that more than half of all gold miners were Indians.

The alliance with Weber held to a point, but most forty-niners did not give a damn whose rights were trampled in pursuit of their golden dreams. While local agreements and partnerships were forged individually, most argonauts agreed with the overall government policy of Native extermination. In order to subsist in this complicated, constantly changing world, José Jesús was forced to adapt to the emerging reality, and the chief modified some of his earlier arrangements, providing miners and a workforce for the gold-seeking throng, and accommodating to the world as it was. It was not capitulation but a desperate struggle for existence in a society dedicated to the removal of his people. It was an astonishing accomplishment that *any* Indians survived in that cauldron of greed, ambition, and violent racism.

And then the rush became a flood. Most gold seekers were not as principled as Weber and José Jesús, and the Indians' lot grew increasingly dismal. Colonel Mason's report chronicled the widespread exploitation of Native miners, and violent incidents such as the one at Murderer's Bar became common. Native groups were routinely dispossessed of their land, their property, and their lives as erstwhile alliances were disregarded and the newcomers took whatever they could. Indian miners were unceremoniously dumped into the worst jobs or banished from the diggings altogether, and the Native populace became exiled in its own land. For his part, Weber honorably fulfilled his agreement and continued to hire Indians as miners and vaqueros, but that was small comfort in the midst of widespread abuse.

As late as 1851, the United States' Commissioner of Indian Affairs reported that José Jesús still commanded more than four thousand Natives in thirty bands stretching from the Calaveras River to the Tuolumne River, but that was likely a much smaller number than in previous years.

When José Jesús was shot by John Huddart in 1849, Charles Maria Weber paid for the chief's medical treatment and transportation, but Weber never saw or heard from José Jesús again. The chief returned to his home at Knights Ferry in Stanislaus County and all but disappeared into history. Although commissioners for the Bureau of Indian Affairs sought his counsel and hoped to negotiate new treaties with José Jesús and his people, the chief's name does not appear on any agreements bargained after that time, and there is no record of him in the 1852 census. Some have speculated that he retired to Yosemite or was an anonymous Gold Rush casualty, of which there were thousands. But many believed he was murdered in 1851, the victim of another gunshot. As historian Albert Hurtado concluded:

> Though we may never learn his fate, the record of his life is a fitting monument to the Indians who survived the gold rush. Indians were victimized; but they were not merely victims. They made choices about their futures based on their sense of history and their standards of justice. Accommodating, working, fighting, hiding out—in a word, surviving—they were the seed for today's California Indians.

Charles Maria Weber admired José Jesús, and as he laid out the grid for his valley town of Stockton, he named one of the main thoroughfares in honor of his ally. José Jesús Street ran through Stockton until the late 1860s, when it was renamed Grant Street in honor of President Ulysses S. Grant.

"A VAST DEAL OF KNAVERY"

THE WARNINGS OF DANIEL WALTON

California's gold launched an avalanche of dreams, offered a sparkling avenue of escape, and promised unforgettable adventure and wondrous opportunity. When the discovery of 1848 was validated by official accounts, anecdotal evidence, and exuberant letters home, a seemingly incurable contagion spread rapidly and relentlessly around the world. Overly optimistic guidebooks, often inaccurate maps, mining primers offering magical techniques, and feverish newspaper stories were the carriers of this optimistic malady—gold fever! As Texas's *Corpus Christi Star* exclaimed in January 1849, "Never did cholera, yellow fever, or any other fell disease rage with half the fury with which gold fever is now sweeping over our land. . . . [It has consumed] communities in one fell swoop, sparing neither age, sex nor condition. It is the rage."

Imagine the reactions as people around the world read these words:

> What seems to you mere fiction is a stern reality. It is not gold in the clouds, or in the sea, or in the center of a rock-ribbed mountain, but in the soil of California, sparkling in the sun and glittering in its streams. It lies on the open plain, in the shadows of the deep ravines, and glows on the summits of the mountains.
>
> Rev. Walter Colton, in a private letter to Secretary of the Navy John Young Mason, September 16, 1848

> The accounts of the abundance of gold in that territory are of such extraordinary character as would scarcely command belief were they not corroborated by the authentic reports of officers in the public service.
>
> President James K. Polk, State of the Union address, December 5, 1848

> The Eldorado of the old Spaniards is discovered at last. We have now the highest official authority for believing in the discovery of vast gold mines in California, and that the discovery is the greatest and most startling, not to say miraculous, that the history of the last five centuries can produce.
>
> *New York Herald*, December 9, 1848

> We are on the brink of an Age of Gold. . . . Whatever else they may lack, our children will not be destitute of gold.
>
> Horace Greeley, publisher of the *New York Tribune*, 1848

> [In California can be found] . . . gold dust, gold nuggets, gold ingots, weighing from one to twenty-four pounds.
>
> *Journal des débats* (Paris), February 14, 1849

> Men here are nearly crazy with the riches forced suddenly into their pockets. . . . The accounts you have seen of the gold region are not overcolored. The gold is positively inexhaustible.
>
> *Memphis Eagle*, May 22, 1849

And yet . . . as a centuries-old proverb memorably reminds us, all that glitters is not gold. Beware!

Although they did not tip the assayer's scale to anything resembling balance, there were among the fanatics a number of sober-minded advisors, stern clergy, and practical friends and family who warned zealous gold seekers to proceed with caution or, better yet, to just stay home.

On an invigorating January morning in 1849, a year after James Marshall had detonated the Gold Rush, forty-two-year-old Rev. Dr. Elisha Cleaveland approached his pulpit at the Third Congregational Church on Court Street in New Haven, Connecticut. His congregation was composed mostly of bright young men attending Yale. Many had been bitten by the gold bug and were seriously contemplating a journey to California. Cleaveland's goal was to dissuade these Bulldogs from the trek. It would be unacceptable, he thundered from the pulpit, for them to join "the filth and scum of society" who "poured in there to seethe and ferment into one putrid mass of unmitigated depravity." These were profoundly powerful words for a generation that equated personal failure with moral deficiencies. Cleaveland hoped to remind them that no good could come to them by associating with those so beneath them. But perhaps Cleaveland set too much store in his message and in his power as messenger, because he continued rather imprudently. The reverend doctor's scolding was immediately undercut by his next words. He attempted to explain that yearnings for gold were only natural given that "within our borders [is] a tract of country larger than New England, underlaid, we know not how deep, with *pure gold*." This bounty is waiting, he continued, "in immense quantities, from the sands of the rivers, or picked from the rocks, or gathered from the surface of the ground . . . free to all as the air we breathe." One can easily imagine these young men leaping from their seats in excitement and leaving immediately for the land of golden dreams.

Perhaps Cleaveland's fire and brimstone was easy to ignore because he, like his parishioners, could only speculate about the goldfields from thousands of miles away. Daniel Walton was different. Or he should have been. He had been to California and collected his thoughts in a slim 1849 pamphlet with the wonderfully cynical title *Wonderful Facts from the Gold Regions*. Walton was unsparing in describing what he viewed as the "latest well-authenticated facts from the gold regions." In his introduction, he expressed the hope that "a perusal of this pamphlet will induce some to 'look before they leap' into a gulf from which they may never extricate themselves." Walton's booklet was widely read, possibly because the title misled its readers into believing the content was an enthusiastic confirmation of riches beyond compare. As with Cleaveland, however, his counsel was also ignored; his stark vision was one the vast majority of gold seekers was not willing to consider.

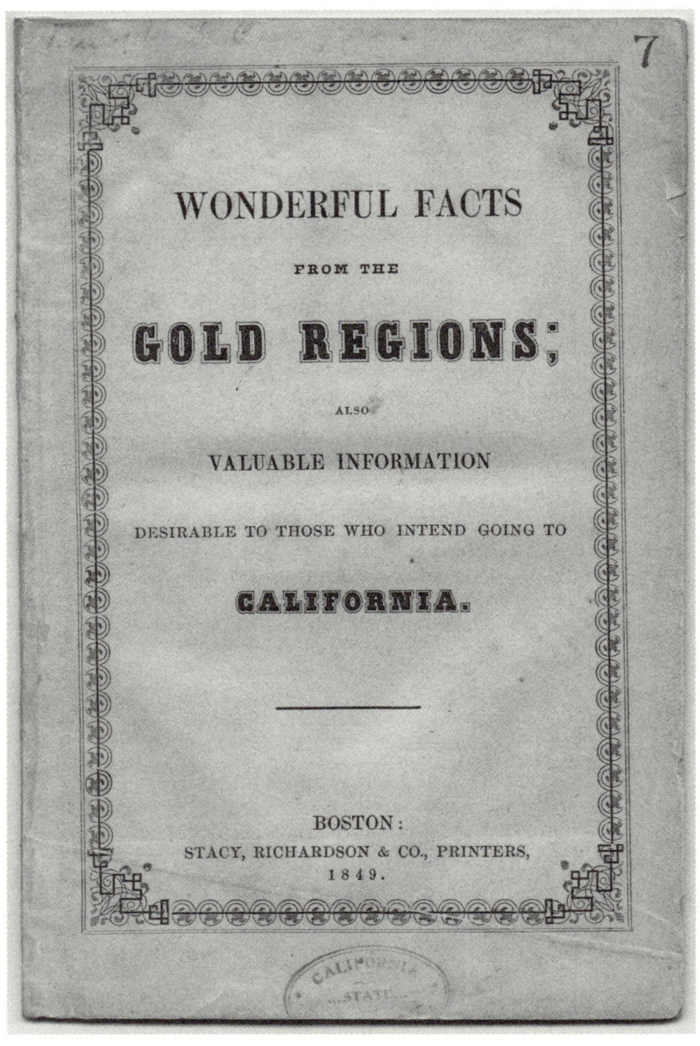
7

WONDERFUL FACTS

FROM THE

GOLD REGIONS;

ALSO

VALUABLE INFORMATION

DESIRABLE TO THOSE WHO INTEND GOING TO

CALIFORNIA.

BOSTON:
STACY, RICHARDSON & CO., PRINTERS,
1849.

Despite his book's optimistic title, Daniel Walton offered warnings instead of unrealistic promises for the gold seekers. Cover of *Wonderful Facts from the Gold Regions*, by Daniel Walton (Boston: Stacy, Richardson and Co., 1849). Courtesy of the California State Library, Sacramento; California History Section.

As it was, Daniel Walton's cautions were brutally unvarnished and repeatedly accurate:

> The words Gold Mines are used with respect to this discovery; but, in truth, thus far *mining* has had very little to do with the matter—digging would be a much more appropriate phrase.
>
> The value of gold is constantly falling.
>
> The markets contain nothing whatsoever and it is with great difficulty that anything, even the common necessaries of life, can be obtained.
>
> The present excitement will attract vast numbers of the idle, vicious, and dissolute.
>
> Let every man . . . who goes to California, arrange his affairs at home, if he has any to arrange, and *make his will*.
>
> We would earnestly admonish to beware of being crazed by the wonderful reports which shall continue to roll in from the "gold diggings."
>
> [Gold] lumps of three pounds! Monstrous nonsense.

And finally this:

> If people believe all the accounts which are published about the gold mines, they will be woefully mistaken. . . . Our own opinion is there is a vast deal of knavery in getting up this gold fever, —and that thousands and tens of thousands will be cheated out of their hard earnings, and reduced to poverty, privations, and even death by "seeking the treasures of the earth."

The forty-niners had been warned. Few listened. The fever was too strong.

"THE EL DORADO OF THEIR MOST SANGUINE WANTS"

JOHN LINVILLE HALL, THE FIRST JOURNAL

The California Gold Rush is one of the best-documented events in history. Seemingly everyone who mined, rode, walked, dug, cursed, caught a cold, had a beer, developed blisters, spit, sold goods, gambled, or entertained in the vicinity wrote about it in a diary, letter, letter sheet, or newspaper article. But the most interesting journal of all may have been the very first one, written by John Linville Hall in 1849.

Hall was from Springfield, Massachusetts, and he caught gold fever early. Before leaving for California, however, he joined something called a joint-stock company—in his case, the Hartford (Connecticut) Union Mining and Trading Company, which was preparing to sail around Cape Horn to the goldfields.

Joint-stock companies, sometimes referred to as "mutual associations," were intricately structured companies founded on the principle of shared work, sacrifice, and profits, and they were especially prevalent in the eastern states in 1849. After President James K. Polk confirmed the "abundance of gold" in California in his December 1848 State of the Union address, joint-stock companies began to form within weeks. On January 24, 1849, the *New York Herald* wrote that forty-seven companies with a combined total of 2,499 members were readying to embark for the golden shore and, on January 29, the

Herald hyperbolically suggested that "ten thousand and one mutual associations . . . have sprung up like mushrooms."

Many joint-stock companies borrowed the concepts expressed by the Perseverance Mining Company of Philadelphia. Company officer Samuel Upham noted that the associates swore to "pursue such business in California, or elsewhere, as shall be agreed upon by a majority of its members, and that the expenses of the company shall be mutually borne and the profits equally divided. . . . We hereby pledge ourselves . . . in all cases to stand by each other as a band of brothers." Frequently the companies would adopt formal articles of agreement that not only detailed divisions of labor but offered solemn moral promises, such as to not drink, gamble, or curse. Members of a joint-stock company were often from the same town, and the cooperative structure created a communal sense of purpose and a feeling of unity even before the enterprise was under way.

Individual members could purchase stock in amounts that ordinarily varied from fifty to several thousand dollars; the more stock you bought, the greater your share of the spoils. The funds raised would purchase and outfit a ship for the sea route around Cape Horn or be used to acquire wagons and oxen if an overland trek were preferred. Company members would fill their ships or wagons with provisions they felt would be sufficient for at least one year, but this guess was almost always wrong, and many an adventurer found himself short of supplies much earlier than anticipated. Just as it was true for other forty-niners, this was an undertaking based almost entirely on rumor and conjecture.

John Linville Hall may not have grasped what awaited him on the other side of the continent, but he committed himself to documenting what happened along the way. Having worked as a publisher, he took his printing press along for the ocean voyage and published a daily account of the journey while on the Hartford Union Mining and Trading Company ship, the steamer *Henry Lee*. These writings were later collected and published as *Journal of the Hartford Union Mining and Trading Company*.

Hall's joint-stock company had 122 members; about a quarter were married, and more than thirty trades and professions were represented, including farmers, machinists, carpenters, merchants, boot- and shoemakers, blacksmiths, clerks, tailors, cabinetmakers, and painters. The company had raised $35,000 in capital and believed they had enough money and provisions to last two years in the goldfields.

In the journal's preface, Hall expressed the company's hopefulness—a sentiment echoed by many later gold rushers:

> California, the *pass-word* and *by-word* with all, has gathered around it a potency which knows no equal. When spoken on the corners of the streets all within reach of its sound bend their heads and listen with silent attention. . . . Many have imagined the rainbow resting its beautiful colors on golden ground, but few thought the sun, moon, and all the starry hosts were day and night pointing the nations of the East to the El Dorado of their most sanguine wants, *gold*.

The entries begin on February 17, 1849, when Hall wrote with optimism: "The sun shone bright on the morning . . . ; a morning welcomed by us all—when we were to bid farewell to all that had been familiar to us and enter upon the novelties of an untried kind of life."

For months the sojourners hopscotched down the Atlantic coastline of North and South America. They arrived at Cape Horn in mid-June 1849, which was wintertime in the Southern Hemisphere. Hall's account of the nearly two-week passage around the cape is one of the most literary, lyrical, and heart-stopping descriptions of the voyage around the southern tip of South America:

> Snow comes here not in the form of flakes as at home, but in *round* particles and often large. . . . Many suffer chilblains and swollen hands. . . . High seas—deep rolling of the ship—rain and snow—head winds . . . a thick, heavy atmosphere rests upon us . . . a sky above all darkened with clouds, and bleak storms daily driving in our faces . . . drenching showers of rain—interrupted by fiercer ones of

JOURNAL

OF THE

HARTFORD UNION MINING AND

TRADING COMPANY.

Containing the name, residence and occupation of each member, with Incidences of the Voyage, &c. &c.

PRINTED BY J. L. HALL,

On board the

Henry Lee.

1849.

Numerous personal journals were written during the Gold Rush, and many believe the first one was by J. Linville Hall. Title page of *Journal of the Hartford Union Mining and Trading Company* (1849), by J. Linville Hall. Courtesy of the California State Library, Sacramento; California History Section.

> snow and sleet. . . . These stormy winds and raging waves continue to be familiar elements, and although previously fond of listening to the wild 'storm fiend' . . . we are not desirous of their longer presence.

Soon after rounding Cape Hope, they encountered another vessel, the *Loo Choo*. A voice from the *Henry Lee* cried out to the captain of the *Loo Choo*, "Any news from California?" The captain responded, "Yes, good news; plenty of gold, no women."

The *Henry Lee* reached San Francisco Bay on September 13, 1849, seven months after leaving the East Coast. Hall excitedly wrote:

> Our approach to American soil acted upon our feelings as upon the exiles at the first glimpse of the heaven-pointing spire of his own village church, and the paternal roof which sheltered him in childhood. What sensations thrill his soul! How animated his countenance! How sparkling his eye! How elastic his step! Yet, seemingly, how slow he passes every object!

John Linville Hall exults in finally standing on terra firma when the company disembarks at San Francisco. However, after "treading the moving, jarring, and tripping deck," Hall and his companions must make adjustments as "he who rides for months upon the ocean waves, finds himself too often, when first upon land, undulating, curtseying, and reeling." And the unsteadiness of dry land after so many months at sea was not their only shock. Upon arrival in San Francisco, the company was disheartened to learn that the goldfields were not waiting at their feet. As Hall noted, "Some expected that they would anchor close to the mines, and make use of the ship as a sleeping place at night; but when the nearest mines to the shore were found to be from fifty to one hundred and fifty miles, and some twice that distance, their illusions were dissipated." This was a common misconception among the early argonauts.

The remainder of the journal is relatively brief, with some tantalizing descriptions of 1849 San Francisco:

> San Francisco is situated about ten miles from the entrance of the harbor. A semi-circle of high hills extends around a little green basin-like centre, which had received the name of *Yerba Buena*. Climbing the hill now called Telegraph hill, we have one of the most inspiring views that falls to the lot of man. . . . Down upon the medley of tongues and nations that compose the city, we now may gaze. Houses and sheds were thrown together, and a multitude of tents, in every shape and size, marked the streets. Not more than 100 wooden houses were built where stood in two short years afterwards upwards of four thousand dwellings, and some of them of costly fabric and architecture.

Hall devotes nearly half of his short section on San Francisco to describing the "inviting houses and tents" designed to "allure the unwary and soothe the disappointed."

> Gold, and silver coin, and ore, were heaped upon the tables, around which gathered the victims of the gambling mania. Men, maddened by their losses, stood at the bar drinking the fiery liquid, to brace their sunken spirits; boys, with bags of dust, and handfuls of coin, were smilingly tempted on, and permitted to win and indulge the hope of being Fortune's favored ones only to be disappointed and ruined in the end.

With high hopes, the Hartford Union Mining and Trading Company members headed to the goldfields to discover their long-sought El Dorado. John Linville Hall and his compatriots had work to do. And then the journal abruptly ends.

More than forty years later, Hall added an appendix to his journal that filled in what had happened. Almost immediately upon arriving in California, he wrote, the company disbanded. The members formed partnerships of two or three people each and headed inland. Hall ended up on the American River and failed in the diggings:

> The placer digging is the poor man's mine, requiring few and simple tools, and hard labor. To handle tons of rock,

> and loose stones and earth in order to find a single grain of gold, as tens of thousands have done, is discouraging to the laborer, who can find better remuneration at other business. The success of one in a hundred is heralded abroad, exciting the multitudes into hope, while the failure of ninety-nine is not printed, or even reported. The ninety-nine generally chose to say nothing, and the world takes no note of their disappointments.

John Linville Hall mined in Weaverville (Trinity County) and Placerville (El Dorado County), but to no avail. He eventually gravitated to Sacramento, where he found that his funds were "reduced to $1.50." He wrote with great sadness: "I sat on the banks of the Sacramento, near where the American River empties into it, dejected in spirit, and thinking of home, of kindred, and wife, and child, that I might never see again."

Calling upon his previous experience back East, Hall became an itinerant printer. After scrambling to safety in the massive Sacramento flood of 1850, Hall traveled to San Francisco and continued plying his trade. On May 3, 1851, a massive fire enveloped the city, destroying one-quarter of San Francisco, killing at least nine, and reducing more than two thousand buildings to smoking rubble. Hall barely escaped the conflagration.

After fire, flood, and failure as a miner, John Linville Hall decided to go home. On May 6, 1851, just three days after the fire, he booked passage on the sailing ship *Florida* for a voyage to Connecticut via Cape Horn. In June, the vessel stopped in Panama for supplies. Hall, reckoning he would never have another opportunity to visit the jungles and mountains of Central and South America, left the ship to explore, and the *Florida* continued on its way. Several weeks later, in the Straits of Magellan, the *Florida* was commandeered by escaped Chilean prisoners. The convicts killed everyone on board.

John Linville Hall eventually returned to Connecticut and, after reflecting on his several close brushes with death, devoted the rest of his long life to religion. He was a respected Protestant clergyman until his death in 1910 at the age of eighty-eight.

"A PERFECT USED UP MAN"

IN THE DIGGINGS

In the earliest months of the California Gold Rush—the fabled days of 1848 and early 1849—there was one thing almost everyone considered a certitude: that there was a limitless golden bounty just waiting in the foothills. Peter Burnett, who became California's first governor, excitedly exclaimed in a May 1849 letter, "THE GOLD IS POSITIVELY INEXHAUSTIBLE." But beyond that one detail, little was known about what kind of life people might find out West. There were questions as to what dangers lurked, what mining techniques were used, what equipment was required, and even where exactly Gold Country was. In those early, rousing days, those who went to the goldfields carried equal measures of zeal and ignorance. In most cases, they did not find El Dorado but rather frustration and disillusionment.

When the news of James Marshall's January 1848 gold strike was widely disseminated, the population grew rapidly and markedly. Despite the difficulties in reaching isolated California, by July the number of immigrants to the diggings was around 2,000, by October the total exceeded 5,000, and by the end of 1848, more than 8,000 newcomers had arrived. This was a drop in the bucket compared to 1849, when an estimated 89,000 people arrived by land and sea, many of them compelled by visions of instant wealth, and with no conception of the hard work that that awaited them. Jonas Winchester, a recent arrival from New York, wrote in a November 19, 1849, letter:

> The incredible difficulties and hardships attending the operations of mining are no more understood . . . than the hieroglyphics of the pyramids. . . . The digging and washing processes are the hardest kind of work a human being ever performed. I have been in the water a week, up to my knees all days and scraping up dirt from the rocky bottom with a big spoon and my hands. The stones and gravel have worn the nails from all my fingers, down to the quick, by scratching on the rocks and picking out stones. And in stooping constantly, it sometimes takes a minute to straighten up. . . . This is a strange world, unlike anything ever dreamed of.

As for the communities that formed, the throng of new arrivals was not interested in settlement. They did not care about stability or establishing durable institutions. Instead, they were inspired by the fantasy of striking it rich. Charged with excitement and bursting with hope, the miners' personal adventures were temporary and unpredictable, and their shelters, clothing, food, and work in the diggings reflected this impermanence. The mining camps that blossomed overnight were rude affairs, a universe of lean-tos and canvas tents. The placer mining communities—as opposed to the towns that sprung up around established underground mines—were either "diggings" or "dry diggings," the distinction being that regular diggings were located on watercourses and dry diggings were not. Whatever the name, these were ephemeral encampments that could multiply from a dozen miners to hundreds within a day. The residents knew that portability was a key characteristic, and as spontaneously as they arose, they could disband just as quickly. If the paydirt played out or a rumor floated through the camp of a richer placer just over the ridge, the miners grabbed their gear and lit out for the new opportunity.

Merchants set up shop by simply slapping a wooden plank on top of two barrels. Gold dust was the currency. Shopkeepers generally did much better than the miners. New Yorker Jared Comstock Brown, a twenty-five-year-old blacksmith in Coloma, El Dorado County, bragged to his father, "I get 8 Dollars for horse shoing[.] I can beat any thing in this country on shoing. . . . There is immense amounts of gold

Despite coming from many different cultures and backgrounds, Gold Rush miners shared many similar experiences in the diggings. "Portrait of 23 Miners at Sluice," daguerreotype by William Chapman, c. 1852. Courtesy of the California State Library, Sacramento; California History Section.

here but it takes a great Deal of labor to get it[.] I have not mined but one Day since I have been here[.] I shall stay in the shop as long as it will pay me."

These boomtowns were overwhelmingly male, and, in most cases, exclusively male. Most had a mixture of hardened veterans and neophytes; an "old miner" was a relative term. John Swan, an English sailor who briefly mined on the American River in 1848, wrote that "a man that had been a month or two in the mines was considered an experienced miner and was often looked on as somewhat of an oracle." The greenhorns were fascinated by the experienced miners. William

Perkins, a hopeful migrant who eventually became a merchant in Sonora, Tuolumne County, recalled his first encounter with the old hands:

> Here were real, live miners, men who had actually dug out the shining metal and who had it in huge buckskin pouches in the pockets of their pantaloons. . . . These men were the awful objects of our curiosity. They were the demi-gods of the dominion. . . . Their long rough boots, red shirts, Mexican hats; their huge, uncombed beards covering half the face . . . all these things were attributes belonging to another race of men than ourselves, and we looked upon them with a certain degree of respect and with a determination soon to be ourselves as little human-like in appearance as they were.

The vast majority of these newcomers were inexperienced, or "green," miners. Many had been farmers or had been toughened by the endless days on the overland trail to California and were used to hard work, but placer mining was unlike any work they had done before, and these novices struggled. As Sheldon Shufelt, a forty-niner from New York, wrote, "We did not have very good success being green at mining, but by practice & observation we soon improved some." An "old timer" named Miffin begged to differ:

> I am amused very much every day at the maiden efforts of the green 'uns . . . at gold hunting. They puff and blow like young whales; their hands soon blister; they bespatter themselves with yellow mire; occasionally they slip up, and souse their seats of humor into a cold bath—at all of which they make all sorts of comical wry faces.

Placer miners used a variety of equipment to hunt gold, and many of these tools were new to the men. In addition to picks and shovels, most used the iron pan, about eighteen inches in diameter with sloping sides about three inches deep. This essential instrument was highly portable and simple to use, with or without water, and with some practice and dexterity, the argonaut could swirl a panful of dirt, sand, gravel, and water (if it was available), and watch the heavier gold dust or flakes sink to the bottom. The pan could quickly determine whether a spot had sufficient gold to warrant further digging. Mexican, Native, and Chinese miners used wooden bowls or baskets that served the same purpose. Many placer miners also used a "rocker" or "cradle," which operated on the same principle as the gold pan but allowed a greater volume of dirt and gravel to be processed at a time. Easy to assemble, the rocker was four to six feet in length. The paydirt was shoveled into the apparatus, which would then be rocked to sort the dirt. Water and lighter materials would pass through the rocker, and the heavier and hopefully gold-bearing particles would be deposited on a metal cleat, called a riffle, at the bottom. In some cases, impatient miners built a "long tom," which was essentially a larger version of a rocker—an inclined twelve- to fifteen-foot wooden trough. Several miners could shovel dirt into the device simultaneously, increasing overall productivity. Adding the force of running water, the clumps of sand and mud would break apart and leave gold on the riffle below.

While there were differences in the practical skills and equipment of individual miners, much else in their lives was very similar overall. Attire was purely functional and seldom washed. As New York miner Prentice Mulford recalled, "Dressing was a short job. A pair of damp overalls, a pair of socks, a pair of shoes, or possibly the heavy rubber mining boots. Flannel shirts we slept in. . . . Vanity of apparel there was little for the working miner. Who was there to dress for?"

Dining options were similarly uniform. In his 1849 book *California As It Is, and As It May Be*, Dr. Felix Wierzbicki described a well-provisioned mining camp as having "pork, bacon, hams, jerked beef, flour, sugar, tea, coffee, chocolate, beans, rice and dried apples, fresh beef and mutton whenever they can get it, which is sometimes the case, and deer meat when they can kill it." But in the transitory world of the placers, delivery of adequate food was tricky. Often meals were quick, greasy, and dreary exercises in eating whatever was available. As gold rusher J. D. Stevenson explained in an April 1849 letter:

> This labor would be more endurable, if at the close of day, [the miner] could enjoy the comforts of good food and rest, but this is out of the question. He must cook his own food,

> or go without it. . . . The food is rarely such as will satisfy the appetite of a fatigued and hungry man. . . . I have seen men living for days without any other food than flour mixed with water formed into a kind of dough and baked in the ashes.

Inevitably, bad nutrition led to ailments ranging from diarrhea to scurvy. As Stevenson noted, "The result is that living in this way produces sickness and disease, and many who come into the town with heavy purses of the precious metal are broken in health and constitution."

The daily drudgery of placer mining was a harsh reality for the "old miners" and "the green 'uns" alike and it had the power to quickly halt any thoughts of sudden fortune. In an 1850 letter to his father in Missouri, forty-niner A. M. Williams wrote:

> A miner's life, I think, is the hardest in the world. . . . He may dig all day, and at night he may have one, two or three dollars—digging, prying away the large rocks, that have been rolling off for centuries from the sides of the mountains, into the stream—while the sun pours down in rays with a power he never before felt—surrounded all the time by strangers who care nothing for him, nor he for them.

For most placer miners, days were an interminable routine of work, fail, eat, sleep, and repeat the following day. Williams described this depressing ritual:

> Night comes on, and he goes to his tree, cooks his supper, spreads his blanket on the ground, and goes to sleep; gets up in the morning out of the dust, cooks his breakfast, and goes to work as before. . . . [He] comes in at night with probably some luck, may be better, may be worse. He works on that way for three or four days, and finding his hole won't pay he hunts another place, . . . finds it won't do—hunts a third, and so on. . . . Such is gold hunting in California.

With no fixed addresses for miners in the camps, mail delivery was irregular, and the men were often as lonely as they were discouraged. Many longed for home. In 1850, E. R. Pratt lamented to his brother in Missouri: "Although I am making money here, . . . I am one of the most unhappy human beings on earth, and shall continue to be till I return to my family. I suppose you know that I always had a roving disposition; you may depend I am cured of it, when I get home again I shall remain there."

A popular Gold Rush song neatly encapsulated the fervor and frustration of the earliest placer miners. It was entitled "Life in California," and it told the story of a Maine farmer who contracts "gold fever." He sells his farm to finance a trip to the beckoning goldfields of California, and leaving behind his wife and children, he sails to the golden shore. In San Francisco, he loses almost all of his stake to a monte dealer but uses what remains to book passage on a paddlewheeler to Sacramento where he "tho't the darned mosquitoes would ha' taken my liver," and then nearly freezes to death while mining in a mountain stream. The unfortunate miner develops a fever and falls very ill. He visits a doctor, who provides a cure in return for the physician's standard fee: all of the miner's gold dust, representing five months' work. Penniless and disheartened, he moans:

> I'm a used up man,
> A perfect used up man,
> And if ever I get home again,
> I'll stay there if I can.

"WITH A TAINT OF FRAUD AND A SPICE OF COMEDY"

CLAIM SALTING

After only seven months in the goldfields, William Swain was an old hand. An emigrant from New York, the twenty-nine-year-old Swain had set up camp on the banks of the Feather River in January 1850. Waiting for him back home were his wife, Sabrina, and his little girl, Eliza. Even through his raging gold fever, Swain remained observant, levelheaded, and careful; in his own words, he "had too much experience to 'go it blind.'" One aspect of mining life particularly troubled him. In a July 19, 1850, letter to his mother, Swain wrote, "There is the greatest swindling, wild and visionary buying and selling in these claims and the gold business generally that ever existed. He who comes into the mines green runs great risk of being outrageously shaved."

The most important thing a miner could possess was his claim, the physical location where the mining occurred. In the earliest days of the Gold Rush, staking a claim was a free-for-all—first come, first served. When a potentially rich gold deposit was discovered, word spread quickly through the diggings grapevine and miners would, as forty-niner Charles Peters reported, "swarm like bees" to the site. With a degree of mutual, voluntary agreement that policed mineral rights and camp decorum, the miners would informally establish mining districts and subdivide and map the areas into claim plats. No uniform standard applied, however, and claim size varied dramatically by location.

Discovery of the rich ground was the paramount factor in those initial weeks, and possession meant control of any individual claim. Technically, all the land was owned by the United States, which freely offered its use for mining, but in the absence of any tangible legal system in those early months, the rules of the mining district became the law of the land. Caustic commentator Hinton Rowan Helper incredulously noted that the government, "in its suicidal liberality, exercises comparatively no jurisdiction" of claims. He also detailed the practical application of claim possession:

> If a man lets his claim go unworked a certain number of days, say five, eight or ten, he forfeits it, and any other person is at liberty to take possession of it. When a miner wishes to quit his claim only for a few days, he stacks his tools upon it, notifies two or three adjoining neighbors of his intention, and goes where he pleases. If he returns within the time prescribed by the laws of the Bar, he is entitled to resume his claim; but if he is absent a day longer, it falls to the first person, without a claim, who may happen to find it.

Miners were supposed to be industrious and work their claims diligently. Idle miners were ostracized. As Theodore Hittell wrote in his 1897 *History of California*: "All men who had or expected to have any standing in the community were required to work with their hands[;] labor was dignified and honorable . . . [and] the man who did not live by actual physical toil was regarded as a sort of social excrescence or parasite."

The cultural diversity of the diggings led to more than a few odd interactions and created an atmosphere ripe with misunderstanding and the potential for fraud. For example, Scottish artist and writer John David Borthwick encountered "one of the greatest curiosities in the country" near Downieville, Sierra County, at a claim populated by an unusual miscellany of nationalities, and headed by a volatile German, whom Borthwick called "the Flying Dutchman":

> The claim . . . consisted of two Americans, two Frenchmen, two Italians, [and] two Mexicans. . . . I passed by [the] claim one day, and such a scene it was! The Tower of Babel was

> not a circumstance to it. The whole of the party were up to their waists in water, in the middle of the river, trying to build a wing-dam. The Americans, the Frenchmen, the Italians, and the Mexicans, were all pulling in different directions . . . while the directing genius, the Flying Dutchman, was rushing about among them, and gesticulating wildly in his endeavors to pacify them, and to explain what was to be done. He spoke all the modern languages at once, occasionally talking Spanish to a Frenchman, and English to the Italians, then cursing his own stupidity in German, and [insulting] them all collectively in a promiscuous jumble of national oaths.

Worse than the indolent were those that shattered the fragile understanding that governed the rather tenuous ownership of claims. As the Gold Rush intensified, claim jumping, or illegally seizing another's claim, became more common. There were also instances of miners being driven from their claims, and Chinese miners were especially frequent targets. Fights broke out and people were murdered. And, in the ephemeral world of the Gold Rush, keeping track of claim possession was challenging at best.

In order to resolve escalating disputes among miners, scores of more formalized self-governed miners' committees crafted rules and regulations specific to their communities and topographies. While there were some similarities—such as equality of opportunity and the rule that possession was marked by leaving one's tools in open sight on the claim—the local laws allowed for wide variations, as in the sizes and utilizations of individual claims. In some densely populated, productive diggings, for instance, a claim could be as small as 15 square feet, while in other locations the claim could be much larger, as was one claim on the Feather River that extended for nearly 660 feet of river frontage. A consistent factor throughout the goldfields, however, was that none of these regulations conferred permanent title to the patch of ground, and claims could be bought or leased. There were also limits on the numbers of claims that a miner could possess in any particular diggings—usually one by discovery and one by purchase.

But one thing that could not be successfully regulated was cheating. "Claim salting" was the term used to describe deliberately and fraudulently altering a claim to make it appear richer than it was. In his 1860 book *Sketches of Travels in South America, Mexico, and California*, Luther Schaeffer recalled:

> I heard of a game successfully practised upon new comers. . . . [Unscrupulous] chaps put about an ounce of gold dust into the bottom of the pit they were digging, and when the next day came round, and with it the strangers, the men who knew there was no gold to be found but what they had placed there, spoke of their claim as yielding rich returns, and as an evidence of it they would pan out *just* a shovel

Possession of a claim was a primary goal for miners, but it could often lead to unscrupulous behavior. Image 1 from *The Miners' Pioneer Ten Commandments of 1849*. Courtesy of the California State Library, Sacramento; California History Section.

> full of dirt, and they might judge for themselves. The result surprised and excited the strangers; they bought the claim for two hundred dollars, threw off their coats, rolled up their sleeves, and went to work, encouraged by bright anticipations and joyful hopes; but as often as they panned out, so often were they doomed to disappointment.

In 1891, Hubert Burgess described an incident from Tuolumne County in 1851, when brazen "salters" trying to unload what they believed to be a worthless claim dramatically doctored a claim in front of the potential purchasers. Briefly diverting the buyers' attention, the deceitful salters tossed a dead gopher snake on the claim and, after histrionically warning all in attendance of a dangerous rattlesnake, immediately destroyed the snake with a shotgun before anyone could see that it was harmless. The blast conveniently created a new hole, which the potential buyers were instructed to pan. And they struck gold—because the shotgun shell that "killed" the snake had been packed with gold dust. The trick worked, although Burgess stated that the claim actually was very profitable.

This was not the only instance of a claim salter being hoisted with his own petard. Theodore Hittell recalled an incident that had "the taint of fraud and a spice of comedy." Near Georgia Slide in El Dorado County, Hittell recounts that

> the owners of a claim there, supposing it had given out, "salted" it liberally with gold-dust and sold it to a party of newcomers for thirty-two hundred dollars. The purchasers commenced working but soon found they had been deceived. They were of course very indignant and made up their minds to stop work and thrash the sellers..

Before they did, however, the newcomers decided to dig just a little bit more, slightly beyond the crevice they had been working. There they found exceedingly rich ore, and, as Hittell recalled, they realized it was "enough to pay for the claim and large profits besides; and the mine continued to pay well for many years afterwards."

CASTLES IN THE AIR
QUARTZ FEVER

At the height of the Gold Rush, curious references began appearing in the newspapers. Something was sweeping throughout the goldfields . . . something ominous. An 1851 report submitted to the United States Senate by the Department of War warned of a growing "state of aurimania," but linked to that was a strain that threatened to be even more serious than gold fever: quartz fever. A Professor Blake cautioned in an 1852 edition of the *Sacramento Union* to beware of those "carried away by the temporary delirium of the quartz fever." In May 1857, the *Sacramento Union* recalled that "in 1851–52, quartz fever prevailed as an epidemic." In the mid-1850s, the fever broke briefly as miners concentrated on placer gold, but by 1857 a Placer County reporter noted that "quartz fever is contagious [and] it is traveling up these mountains very fast." A reporter from Shaw's Flat, Tuolumne County, echoed earlier accounts, writing that "quartz fever was raging furiously" and "it had become epidemic, and . . . will terminate fatally to your correspondent. It numbers its victims in the hundreds already."

This outbreak—tied to the pandemic of gold fever that animated the entire era—was among the most notable manias, but it was far from the only one. It was just the latest target of the schemes and scams designed to separate the hopeful and naive gold seekers from their money. While mining gold visible in quartz outcroppings was prevalent from the opening days of the rush, this new enthusiasm for the quartz itself was something different, and it revolved around the idea that the quartz could be cooked to yield the gold hidden inside it. While an intriguing idea at first, it eventually evolved into a bamboozle of monumental proportions and consequence.

Many journals from the Days of '49 recall mining companies having purchased "gold-saving machines" before departing from the East Coast. Charles W. Haskins, a forty-niner from Massachusetts, remembered the apparatus in his 1890 reminiscence of his journey with a mining company to the goldfields entitled *The Argonauts of California*. The machines, which were of various sizes and extravagance, all had the same purpose: to extract gold from sand, dirt, and gravel. Possessing such fanciful names as "Bruce's Hydro-Centrifugal Chrysolyte" or "Buffum's Eldorado and Scientific Gold Sifter, Separator, and Safe Depositor," the contraptions promised to save miners from the hardship of panning and digging by hand. A miner would shovel the prospective paydirt into the machine, add water, and turn a crank, and the machine would do all the rest—not only obtaining the gold but also packaging it into bottles. As Haskins described their differences, some "were to be worked by a crank; others, more pretentious, having two cranks; whilst another patent gold washer, more economical and efficient, worked with a treadle." The machines could be operated by a worker standing upright or luxuriating in an attached armchair. The device that accompanied Haskins's company on the journey west required "special mention." He wrote: "This immense machine . . . was in the shape of a huge fanning mill, with sieves properly arranged for assorting the gold ready for bottling. All chunks too large for the bottle would be consigned to the pork barrel." Haskins noted that there was a lively debate amongst the members of the mining company as to whether they were bringing enough barrels to contain all the gold they were assured of finding.

The problem with these expensive "gold-saving" gizmos was that they did not work. Much sweat and drudgery resulted in no riches, just a box full of mud. Charles Haskins describes what greeted him upon landing in San Francisco in 1849:

> We found lying upon the sand and half buried in the mud, hundreds of similar machines, bearing silent witness at once

> to the value of our gold-saving machinery, without the necessity of a trial. Of course ours were also deposited carefully and tenderly upon the sandy beach, from where, in a short time, they were washed into deep water, making amusement for the shrimps, clams, and crabs.

In 1856, Sir Henry Huntley mentioned another California mining swindle, this one requiring investment in a diving bell that would allow underwater mining, thereby avoiding all the fuss and mess of dirt and mud. And an advertisement in the *Palladium* newspaper of Richmond, Indiana, on February 7, 1849, offered "California Gold Grease" to those heading to the region. It reads: "A Yankee down East has invented this specific for the use of gold hunters. The operator is to grease himself well, lay down on the top of a hill, and then roll to the bottom. The gold, and 'nothing else,' will stick to him. Price $10 per box."

Everywhere you went there were these golden visions of endless wealth, wild imaginings of easily accessible fortunes requiring little effort. It fostered hope—oftentimes unreasonable hope—and sometimes the greatest dreams become the worst nightmares. That's exactly what happened in 1851 for investors in Grass Valley and Nevada City.

Following James Marshall's gold discovery in 1848, the earliest fortune hunters were placer miners. Placer mining was the extraction of gold from rich surface gravel in or near watercourses using simple tools such as pans, long toms, or sluice boxes. Some miners went a step further and burrowed small, crude mines into slopes marked by heaps of dirt and gravel deposited at their entrances. These mines were called "coyote holes" and were so common that Nevada City was jokingly referred to as "Coyoteville" in its formative years. But it was the 1850 discovery of precious gold veins in quartz outcroppings that increased the enterprise to a fever pitch in the area.

By 1851, Nevada City had become obsessed with quartz. Placer mining was exhibiting signs of depletion, and miners now turned their attention to standard quartz mining with growing enthusiasm. But quartz mining was expensive. As Prussian-born placer miner Frank Lecouvreur explained from the Yuba River in February 1852: "You know that often quartz is found containing free gold. To get that out the rock is reduced to powder in the so-called quartz mill, and out of that powder the gold is afterwards extracted. You can imagine that such a quartz mill is expensive; the outlay for a small one is about twenty thousand dollars." In today's dollars, $20,000 would be about $600,000.

Capital-intensive quartz mining companies designed to reap this golden harvest blossomed like wildflowers. As the *Sacramento Union* reported in 1857:

> Men totally ignorant of quartz mining became quartz miners; ledges were taken up, mills on a magnificent scale, in the way of expense, were erected by the dozen; machinery of great power, and at immense cost, was sent in every direction into the mountains—every new machine for crushing

THE MINER'S PROGRESS. 11

The Pilgrim rejoiceth ober his "Pile."

But morn dispels these fairy scenes,
And *want* arouses pluck;
He shoulders pick and pan once more,
Again to try his luck.
He digs in dark, secluded depths,
The spots where *slugs* abound,
And oh! what raptures fill his breast—
His "pile" at last is found.

Striking it rich was always a reason to celebrate, but the trail to instant wealth could be strewn with uncertainty, and occasionally deception. "The Pilgrim Rejoiceth over his 'Pile,'" illustration by Charles Christian Nahl, from *The Miner's Progress; or, Scenes in the Life of a California Miner* (1853), by Alonzo Delano and Charles Christian Nahl. Courtesy of the California State Library, Sacramento; California History Section.

> the rock and amalgamating the gold, was tested, and hundreds of thousands of dollars expended which proved a total loss to those for whom it was invested. . . . Those who went into quartz mining with the highest hopes and wildest expectations, came out poorer, if not wiser men.

Often, the only individuals who profited were charlatans fleecing gullible stockholders by offering simpler, cheaper, and less-labor-intensive practices. It is likely that one of these was Professor Rodgers and his Bunker Hill Mine Company on Deer Creek in Nevada City. Professor Rodgers had been a prominent participant in the Quartz Mining Convention held in Sacramento in 1851, and he planned to use an experimental technique of his own devising at the Bunker Hill Quartz Mine. It is unclear if the good professor actually believed in his technique or was just a silver-tongued flimflam artist. In 1867, *Bean's History and Directory of Nevada County* described Rodgers's "discovery" in scientific terms:

> [Rodgers] maintained that quartz was of a porous or cellular structure, but that the interstices between the crystals were not large enough in the natural state to allow the particles of gold to drop out. By the expansion of heat the pores were opened and the metal had free egress either in its cooled or melted form.

In other words, the professor claimed that gold would appear by "roasting" the quartz in an immense furnace. Many leading citizens of Nevada County invested heavily in Rodgers's mine. Investors poured thousands of dollars into the operation, which was advertised as a sure thing.

The first attempt failed. But Rodgers insisted that his idea would work if given another chance. More thousands were invested. As *Bean's History* recounted:

> A large chimney, or furnace, was constructed at great expense, a mammoth wheel erected. . . . Wood and coal in large quantities were procured. . . . The millionaires [investors],

> in expectancy, were on hand night and day, for who can sleep when such a princely fortune is to be harvested? The *savant* who was testing his discovery on a large scale, for a snug salary, rode up occasionally and gave his orders with the air of a General of Division. His employers bowed obsequiously and obeyed his high behests.

Tons of quartz were fed into the blazing inferno, and the investors waited and waited for the gold to arrive. It did not. Speculating the gold was merely stuck on the bottom, the furnace was scraped. No gold. "The bubble burst," *Bean's History* reported. "Dr. Rodgers left the place, and so did a great many others in complete disgust. Quartz was pronounced a humbug." Investors in Nevada City were stunned and embarrassed. They seldom spoke of the event again. *Bean's History* remembered that "houses were deserted, clap-boards hung dangling by one nail, and men went about the comparatively lonely streets congratulating themselves that they were not so poor as to own property in such a doomed city."

In 1856, Nevada City newspaper editor Aaron Augustus Sargent (who would later become a United States senator from California) described the Bunker Hill Mine Company investors as "hundreds [who] made themselves poor by misapplied capital." He explained that Professor Rodgers believed the experiment had failed not through "intrinsic defects in its philosophy" but through poor-quality quartz deficient in gold. The mine's quartz was then crushed by stamp mills in the traditional manner. The outcome? The Bunker Hill Mine had good-quality quartz, but no gold was ever found. The company's lost investment was estimated at $85,000 in 1851—$2.6 million today. Deer Creek was dotted with abandoned equipment from the failed operation. Sargent called it "a huge monument of the fortunes buried there."

Despite having fallen for an enormous boondoggle, residents were able to laugh it off, to a degree. In October 1851, not long after the scope of Rodgers's failure became apparent, Aaron Sargent printed in his newspaper, the *Nevada Daily Journal*, what he called "an amusing burlesque upon quartz operations." It was the annual report of the fictional "Munchausen Quartz Rock Mining and Crushing Company." Munchausen was an obvious reference to the popular literary character Baron Munchausen, who told wildly exaggerated stories and made ridiculous, grandiose claims. Clearly the purported report was satirical commentary on the failed Bunker Hill experiment and is a classic example of the "California humor" genre that originated during the Gold Rush.

According to Sargent, the report featured a mix of actual or slightly altered names of real investors in the Bunker Hill Quartz Mine, in addition to invented characters, such as General Napoleon B. Gulliver, J. Squander Swartwout, Dr. Diabolus Pillgarlick, and Triptolemus Middlefunk. A few well-known figures surfaced as company officers: P. T. Barnum, Guy Fawkes, and Robinson Crusoe. The prospectus specified that the company claimed 405 claims of sixty feet each, "beginning at a blazed dogwood tree on the right bank of the river Styx." The statement further detailed how the miners "are enabled to carry on their work by the light of diamonds, which brilliantly illumine their vast excavations." The report concluded that the Munchausen Quartz Company was characterized by "vague uncertainty and transcendental obscurity . . . [,] sacrificing wealth and enterprise upon the shrine of cupidity"—an apt representation of the Bunker Hill Quartz Mine as well.

Quartz fever did not die with the 1851 Bunker Hill fiasco, it simply transformed into a more profitable enterprise. By the mid-1850s, it was increasingly evident that gold was locked deep underground in extensive subterranean quartz veins, crystalline tributaries streaked with golden filaments. Intensive and expensive underground mining of this "hard rock" produced tons of gold-bearing ore, which was then crushed, or "stamped," in the conventional method and processed into bullion. With this realization, quartz fever returned with a vengeance. Grass Valley and Nevada City, which had seemed fatally stricken by the failure of quartz roasting, rebounded dramatically. The enormous wealth of the Northern Mines was built on the foundation of decades of hard rock mining centered in these neighboring cities, once described as "the nursery of quartz mining in California."

"DEATH STARED THEM FULL IN THE FACE"

SILAS WESTON AND KELLY'S BAR

Silas Weston was far from your typical Gold Rush emigrant. A forty-eight-year-old schoolteacher from Providence, Rhode Island, he stood six foot four and his ancestors had been on the *Mayflower*. Like many before him, he succumbed to the widespread malady known as gold fever in 1852, leaving his home and family behind. He spent eight months traveling to California aboard the ship *Perseverance* and devoted only half as much time to the goldfields—a memorable four months in 1853.

The following year, Weston published his account of the brief visit, entitled *Life in the Mountains; or, Four Months in the Mines of California*. In the event-filled memoir, Weston compiled the narrative from his handwritten notes, composed in makeshift conditions common to the time and place; his "scribbling-desk" was "usually a small trunk, bottom upwards, which he held in his lap while seated upon a stone or the ground." His chronicle begins with wonderment and jovial camaraderie that quickly leads to disappointment and heartfelt longing for home. But it is most notable for the description of his fifth day in the goldfields, which captured a horrifying incident that took place at Kelly's Bar on the American River.

Weston reached Sacramento City on April 10, 1853, and immediately prepared for his journey to Auburn, about forty miles to the northeast. His company consisted of eleven other eager gold seekers who had heard rumors of rich diggings on the American River. Arriving in Auburn on April 13, they learned of a near-fatal Indian attack on a teamster two days earlier. The Indian had escaped amid a flurry of bullets and echoing epithets. Retaliation for the attack had become an obsession in the mining camp.

The next morning, despite serious apprehension and concern for their lives, Weston and a friend decided to visit Kelly's Bar, a supposedly rich mining site outside of Auburn on the Middle Fork of the American River. They had waited months for the opportunity to strike it rich, and they were not about to let a little unrest delay them any longer. Informed by camp veterans that the Indians were, as Weston noted, "exceedingly afraid of the rifle," he and his companion were heavily armed, with two rifles and two six-barreled revolvers. "Our hearts palpitated faster than usual," Weston recalled, "as well may be supposed, as from time to time we anxiously cast our eyes behind the trees, clusters of bushes and hillocks, while passing along, not knowing but the savages would suddenly attack us as they had others."

They arrived at Kelly's Bar—merely a tiny collection of tents on the river—that evening. They soon discovered that the village of the Native who had attacked the teamster was located about five miles up the river canyon and that an expedition had been outfitted to exact revenge. The party consisted of "twenty-two young men, all well armed each with a bowie knife, a short gun [shotgun] and a pair of six barreled revolving pistols, so that the company could fire more than two hundred times without stopping to re-load."

At dawn the next morning, the company headed toward the Indian village for what Silas called their "destructive errand." Ten were on muleback, twelve were on foot.

"Poor Indians!" wrote Weston. "Little did they dream this morning, as the sun arose and greeted them of the awful storm that was about to burst upon them!" The camp suspended mining operations that morning and anxiously waited for the expedition's homecoming and the results of the raid.

Early that afternoon, three of the party returned bearing gruesome mementos. "As they came in, each held in his hand trophies of victory

Forty-niner Silas Weston encountered hundreds of miners, like these two, during his short visit to the goldfields. "Two Miners," daguerreotype, c. 1850. Courtesy of the California State Library, Sacramento; California History Section.

consisting of a small bundle of arrows, from the ends of which hung dangling down, two or three scalps."

Details of the engagement emerged. The Indians occupied a ravine in the forest and had been eating their morning meal. The raiding party prepared their weapons and attacked over the top of a hill, and in response the Indian warriors directed arrow after arrow at the assailants. Weston wrote that the expedition participants stated that "so nimbly did [the Indians] use their bows, that often each had two arrows on the way at the same time, while our men were pouring upon them a most deadly fire."

But the battle was a mismatch. The raiding party had the advantage of more-powerful arms and greater distance; the Natives were simply too far away from their attackers and most of their arrows fell short. Shot and bullets rained on the Native village for thirty minutes. Thirty Indians died, and almost as many were wounded. The raiders suffered two minor injuries. The expedition fired indiscriminately: "Women and children were in the rear, consequently some were killed and wounded—their shrieks and groans were heard by our men as their balls struck them."

The survivors fought valiantly. "The Indians displayed great courage and did not yield, though death stared them full in the face, until their chief had fallen, [at which point] they ceased, and retreated up the hill on the opposite side of the ravine," wrote Weston. Then, "Our men pursued them a short distance and shot several, causing them to fall backwards down the hill. Then with their bowie knives they passed among the wounded and slain, and where life had not become extinct, they extinguished it!" The remaining women and children were forced to beg for their lives.

Weston noted that the waiting miners at Kelly's Bar found the actions of the raiding company to be reprehensible but understandable. "I am happy to say that the miners very generally disapprove of the conduct of the young victors," wrote Weston, "especially those acts of cruelty perpetrated upon the wounded after the enemy had fled. But . . . the Indians have committed so many depredations, that the miners have become strongly incensed against them."

Following this bloody incident, the miners went back to work. Weston and his traveling companion prospected at Kelly's Bar for less than three hours, found nothing, and returned to Auburn late that afternoon.

"THAT BLIGHTING CURSE"
DISSIPATION

Some Gold Rush critics felt that even the relatively tame activities of these independent, footloose gold seekers were wicked, each day placing them one step closer to eternal damnation. Add drinking, gambling, and sexual license to the equation, and the flames of Hell were licking at their bootstraps. Moralists insisted that sin was always lurking just around the corner during the Gold Rush, and it adopted many guises. It could manifest itself as an innocent interlude turned sour, a deeply regretted lapse in judgment, or a dreadful misstep that turned out to be unforgivable. Yielding to desire, they warned, could provide momentary gratification but also empty your pocketbook, alter your reputation, or even cost a life. Some feared that an entire generation of young men would fall victim to the glittery allure of instant satisfaction and be consumed by a whirlpool of depravity. This was an exaggerated likelihood, but many high-minded observers remarked that California Gold Rush behaviors were far from the societal norm and worthy of profound concern. In 1850, Frank Marryat, an English immigrant, wrote that being in California was to be in "the midst of riot, dissipation, and ungodliness." A few commentators contended that yielding to temptation was inevitable—the cost of doing business in the goldfields. North Carolinian Hinton Rowan Helper, who felt that California possessed "the best bad things that are obtainable in America," had an interesting approach to the circumstances. He wrote in 1855 that "no one here can be successful unless he assimilates himself to the people;

he must carouse with villains, attend Sunday horse-races and bull-fights, and adapt himself to every species of depravity and dissipation."

But the majority of critics agreed with New York forty-niner John Letts, who bemoaned the personal transformations wrought by unbridled dissolution:

> Here were young men, who, a few months previous, had left their friends and homes with vigorous constitutions, and characters unblemished, to seek their fortunes in this land of gold. A few short months had sufficed to accomplish the work of ruin. In an unguarded moment they were tempted from the path of rectitude; they visited the gaming-tables and halls of dissipation; and when the brief dream was over, they awoke and found ruin, like a demon, staring them in the face.

The word used most often in journals to encapsulate the self-indulgent debauchery of the argonauts was "dissipation," although there was no consensus on its causes or solutions. Some felt that dissipation was a reaction to failure, loneliness, or despair. Others argued that it resulted from simply having too much time on your hands, coupled with laziness or immaturity. Luzena Stanley Wilson speculated in her memoir of Gold Rush life that "men plunged wildly into every mode of dissipation to drown the homesickness so often gnawing at their hearts." But Englishman William Shaw had a less complicated reason: in 1851 he wrote that young men dissipated because "idleness is the root of all evil."

The motivations for dissipation might be varied and enigmatic, but the form it took was not. Overindulgence in drinking and gambling was the yardstick. Gluttony, profanity, and sexual abandon might become equally prevalent, but whiskey and wagering were the foundation. As Shaw observed:

> The few fortunate diggers would . . . be seen staking their gold dust on cards; gambling more deeply as they became excited, and invariably losing their all, if they continued playing. Others, seated on rough benches, might be seen

> breaking off the necks of champagne bottles; for if they had been fortunate, they took care to show it by ordering the most expensive beverages.

Additionally, Shaw went on, these idle miners might partake of "sardines, turtle-soup, lobsters, fruits, and other luxuries, . . . but the consumers paid very dearly for such epicureanism."

Alonzo Delano, a forty-two-year-old gold seeker from Illinois, recalled that the trigger for dissipation could be subtle. He remembered a mining camp on the North Fork of the Feather River in 1850 that was diligent and decent until a monte dealer arrived one day. Delano noted that the change occurred almost immediately, as if a switch had been flipped:

> Several industrious men had commenced drinking, and after the monte bank was set up, it seemed as if the long smothered fire burst forth into a flame. Labor, with few exceptions, seemed suspended, and a great many miners spent their time in riot and debauchery. Some scarcely ate their meals, some would not go to their cabins, but building large fires, would lay down, exposed to the frost; and one night, in the rain.

Delano recalled that the gambling continued for two weeks. He wrote that "it seemed as if the gold blistered their fingers, and they began a career of drinking and gambling, until it was gone." One miner lost $900, another $800, "their whole summer's work—and went off poor and penniless." The monte dealer won $3,000 in two nights and departed for another camp six miles distant, where he lost $4,000 on his first day in the new location, "exemplifying the fact, that a gambler may be rich to-day, and a beggar to-morrow."

Reporters noted that dissipation seemed to follow a predictable pattern. The caustic gold seeker John David Borthwick, who possessed a particularly mordant sense of humor, described the blueprint at length. The "loafers," wrote the Scotsman, worked only one day a week and then

> spent the rest of their time in bar-rooms, playing cards and drinking whisky. . . . They were always in debt for their board and their whisky at the boarding-house where they lived; and when hard pressed to pay up, they would hire out for a day or two to make enough for their immediate wants, and then return to loaf away their existence in a bar-room. . . . I never, in any part of the mines, was in a store or boarding-house that was not haunted by some men of this sort.

Others "with more energy in their dissipation," Borthwick continued, would work hard in the mines, save their gold dust, and then "would set to work to get rid of it as fast as possible." The individual reveler "went about it most systematically." First, he would buy a new suit of clothes. Then he would visit all the saloons in the neighborhood and, as Borthwick noted, "insist on every one he found there drinking with him, informing them at the same time (though it was quite unnecessary, for the fact was very evident) that he was 'on the spree.'" Soon, the young celebrator would be drunk, but

> before being very far gone he would lose the greater part of his money to the gamblers. Cursing his bad luck, he would then console himself with a rapid succession of "drinks," pick a quarrel with some one who was not interfering with him, get a licking, and be ultimately rolled into a corner to enjoy the more passive phase of his debauch.

After a fitful night's sleep, Borthwick observed that the lightheaded carouser would

> not give himself time to get sober, but would go at it again, and keep at it for a week—most affectionately and confidentially drunk in the forenoon, fighting drunk in the afternoon, and dead-drunk at night. The next week he would get gradually sober, and, recovering his senses, would return to his work without a cent in his pocket, but quite contented and happy, with his mind relieved at having had what he considered a good spree.

The idle behavior of the forty-niners provided a rich lode of material for observers, including Mallorcan artist Augusto Ferran. “Posiciones Comodas” (“Comfort”), illustration by Augusto Ferran, from *Album Californiano* (Havana: 1849–50). Courtesy of the California State Library, Sacramento; California History Section.

Social reformer Eliza Farnham was particularly alarmed by the prospect of dissipation not only for young men but for the handful of children in the area as well. In her 1856 book *California In-doors and Out*, she wrote:

> Those who are disposed to indulgence take full license. Drunkenness, carousing, profanity, are frightfully prevalent. The children participate in all the vices of their elders. I saw boys, from six upward, swaggering through the streets, begirt with scarlet sash, in exuberant collar and bosom, segar in mouth, uttering huge oaths.

Reformers believed that the greater the population, the less likely the impulse toward intemperance; the unruly, they figured, would be outnumbered and dominated by those seeking civility and stability. In their 1855 *Annals of San Francisco*, Frank Soulé, John Gihon, and James Nisbet dispelled this notion. They noted that as San Francisco matured during the Gold Rush there were indeed more churches, more charitable organizations, and more civic improvement societies, but also "more places of dissipation and amusement, more tippling and swearing, more drunkenness and personal outrages, nearly as much public gambling and more private play." The authors wrote that calls for gambling and saloon reform received "a fresh coat of paint" but that debauchery continued largely undiminished. In California, they commented, the natural impulse toward civilization lagged behind those "devout worshippers of mammon" who defined society, a "strange mixture of peoples, . . . with many scoundrels, rowdies, . . . sharpers, and few honest folk. . . . In the scramble for wealth, few had consciences much purer than their neighbors[;] few hands were much cleaner." And besides, they added, "the vices and foibles, the general mode of living, that frightened and shocked . . . at first, seem natural to the climate, and, after all, are by no means so very disagreeable."

For many of these young adventurers, the only effective answer was to remove themselves, however reluctantly, from the temptations of California and return to families waiting anxiously back home. As nineteen-year-old Wisconsin forty-niner Lucius Fairchild wrote to his father on March 17, 1850, "No person should throw up a good business and come to this country. . . . I would never run the risk if I was again at home and it is said by every body, almost, . . . 'if I ever get back I will stay & God forgive me.' . . . If God will forgive me, and allow me to make a small pile this summer I never will trouble this country any more, but break for home. I get home often in my dreams."

During the 1880s, James Steele, a magazine editor and former army officer, visited California and encountered some former gold rushers. In his 1889 book *Old Californian Days,* Steele was struck by the fact that these men continued to be influenced by "an experience now impossible in any corner of the world." For most, Steele wrote, the Gold Rush was a pleasurable memory rife with recollections of youthful indiscretions and days of dissipation. These self-indulgent episodes were usually momentary lapses, however, and, as James Steele noted, "Many a deacon in good standing now was not so then." But Steele was saddened by those who never left their Gold Rush recklessness behind. These "Argonautic relics" were men "whose experiences are only wide, not deep," who wished "to do all those things now . . . which a man should do only in his youth, if at all."

James Steele's description was an echo of James Letts, who wrote of the dissolute in 1853: "They had neither means nor character, and their constitutions had been laid waste by the blighting hand of dissipation. Who can calculate the hours of anguish, or tears of blood that have been wrung from the hearts of bereaved parents and friends by that blighting curse."

"PLENTY OF JABBERING AND QUARRELING AND SEVERAL FIGHTS"

ALFRED DOTEN

Alfred Doten didn't give a damn. For fifty years he kept a private record of every disreputable episode, sleazy encounter, angry confrontation, fistfight, and fandango in his alcohol-fueled life. He was rebellious, bigoted, an epic carouser, and a pompous ass. He was a roguish vagabond, an entertaining raconteur, a bloviating provocateur, a meticulous chronicler of his era, and an occasionally lyrical wordsmith. His perspective was often through the glistening bottom of a whiskey glass or blurred by a black eye, but Doten's descriptions of the California Gold Rush may be the most forthright and revealing ever documented.

Alfred Doten came into this world in 1829, the pious son of a New England seafarer and a direct descendent of the Pilgrims. The sober young man worked as a carpenter and a fisherman, but his father, Samuel, constantly worried about his son's future. Samuel felt that Alfred lacked ambition and the steadfast work ethic of his Puritan ancestors and something must be done about it. Suddenly, a solution arose. The hometown newspaper, the *Plymouth Rock*, published, on February 1, 1849, a letter from former resident James H. Gleason describing the extraordinary opportunity waiting in his adopted territory of California. Gleason wrote, "A very rich gold mine has lately been discovered about 250 miles north of this place [Monterey], and everybody in the country is rapidly hastening towards it. . . . Persons of my acquaintance who could only show $100 two months since, can now count their ten or fifteen thousands."

Gold fever struck and, within a week, the *Plymouth Rock* listed Alfred Doten as a passenger on the converted whaling bark *Yeoman*. Alfred was a member of a hastily arranged association called the Pilgrim Mining Company. Under the terms of the company's charter, free passage to San Francisco would be provided in return for two-thirds of mining profits returning to the company investors. Samuel Doten was an investor and part owner of the *Yeoman*; it is likely that Samuel contrived to send his son to California.

After a seven-month journey around Cape Horn, the *Yeoman* arrived in San Francisco and straight-laced Alfred Doten celebrated by singing temperance songs aboard ship with his compatriots. But then he set foot on the golden shore and all hell broke loose. Exposed to a gaudy, intoxicating vision of glitter and excess, Alfred was never the same again. In a letter to his father on October 6, 1849, from San Francisco, Doten is wide-eyed with wonderment:

> Our company have resolved to stick together, and we start up the river tomorrow, but how long we shall hold together is uncertain, for Gold! Gold! Gold! turns the heads of the wisest. . . . I wish I could begin to give you an idea of this place. It looks just like a muster field, only a great deal more so. There are very few women here, but a plenty of liquor of all kinds, cheap; every house nearly is a gambling house, and gambling is a common pastime. Fortunes are lost and won in five minutes. 36,000 dollars was risked upon the turn of a single card, and lost. I have seen men come tottering from the mines with broken constitutions, but with plenty of the "dust," and sitting down at the gaming table, in ten minutes not be worth a cent. Money is nothing here; the tables groan under millions in gold and silver. But do not suppose there is no law; Lynch law prevails here—Just before we came in, two men were hung for stealing two hundred dollars, and a little boy had his ears cropped close to his head, for stealing 1400 dollars.

Doten and company headed to the goldfields near the boomtown of Sonora. Mostly, they failed, and the Pilgrim Mining Company was shredded by discord. Within a few weeks, the company disbanded and the investment was lost. Alfred Doten was now independent but penniless, and his New England piety dissipated as quickly as his prospects had. His life became a ceaseless string of fighting, women, singing, dancing, cock-fighting, bear-baiting, and drinking—lots of drinking.

He joined a group of coarse, leathery miners in Calaveras County who panned gold by day and spent most nights lost in an alcoholic stupor. They drank so much that Doten soon had to invent new ways to catalog and categorize his drunkenness. In his journal he called nights of drinking a "tall spree" or a "small bit of a spree" or a "soiree" or a "bender" or a "jollification" or a "corrective for a stomach" or a "jubilee." He didn't just get drunk, but "tight" or "less tight" or "a little tight" or "very tight," and sometimes "obscure" or "very obscure indeed" or "somewhat obscure" or "a little obscure."

The formerly virtuous descendent of the Puritans also became obsessed with sex. In his July 21, 1850, journal entry celebrating his birthday, Doten wrote: "My birth day today—21 years old today—Hurrah old man how's your *crotch rope* . . ." He pursued numerous sexual liaisons in the goldfields, usually with women of color, married or not:

September 20, 1852

> This forenoon two squaws came over from the Rancheria and paid me quite a visit—One of them was Pacheco's wife . . . [;] she was accompanied by an old hag of a squaw. . . . I gave [Pacheco's wife] several presents and made myself quite thick with her and after a while I got her . . . and took her into my tent and . . . was about to lay here altogether but the damned old squaw came in mad as a hatter and gave the young gal a devil of a blowing up—Nevertheless I still left my hand in her bosom and kissed her again right before the old woman. . . . I told [the old woman] to go to the devil for her pains and spent some minutes cussing her in good round English.

May 21, 1853

> "Maria" has fell in love with me and wants to have me for her lover as she is a yellow gal and it would hardly do for me to marry her legally—She is worth 15.00 dollars—She is a real good looking girl of fine shape and no doubt a fine bedfellow.

To make ends meet, Alfred opened a store and dance hall at Lower Bar on the Mokelumne River in Calaveras County. It was exactly the kind of establishment one might expect to be run by a hard-drinking carouser. On June 13, 1852, his journal described a typical Sunday in the establishment:

> Today I was full of business as a dog is full of fleas—All day the store was full of drunken Chilenos, French &c and the day passed off finely with plenty of jabbering and quarreling and several fights in which some eyes were blackened and noses bled—but no one was hurt very bad.

A few weeks later, on August 13, 1852, Doten noted one more representative afternoon: "Passed as usual—Trade brisk &c—Had a small bit of a scrape in the evening with Bob Paine, he being drunk, and as he darned and insulted me somewhat I made him see stars and put him out of doors."

Alfred Doten's dissolute routine continued unabated and was the focus of his diary, but he memorably commented on other incidents as well. Doten was hedonistic, but he had a penchant for observation and reportage that would later prove useful. He chronicled the hanging of two criminals at Rich Gulch, near Mokelumne Hill, in 1851:

> After they had hung about half an hour, we left them to swing in the wind till morning. . . . One hardly moved after he was run up and seemed to die easy, but the other one writhed about and seemed to die hard—The taking of the murderers, the trial and execution was carried on in the most quiet and orderly manner throughout—The night was

Alfred Doten, Gold Rush diarist and ne'er-do-well. University of Nevada, Reno, Doten Collection, UNRS-PO189-3.TIF.

> dark and fearful and together with the howling and roaring of the wind through the tall pines and the warring of the elements rendered the scene awful and terrific in the extreme and one that will never be effaced from the memory of those who witnessed it.

Doten also addressed a subject that surfaced in accounts of the time—suicide in Gold Country—but his version was unusually detailed compared to the succinct descriptions in other narratives. An English gold seeker named George had seduced a married woman in New Bedford, Massachusetts, and, Doten wrote,

> probably the thoughts of the wrong he . . . committed was one of the causes of his committing suicide. He has also been sick with the fever and ague several weeks and having also taken to drinking, he had become very much depressed in spirits, and had once before attempted suicide by shooting himself with a shot gun but was prevented—He shot himself with a pistol. The ball passed into the right temple, entering the brain, thus causing instant death—George was a good fellow—We were great friends—Heaven rest his soul.

As for Doten, his alcoholic adventures in the Mother Lode came to a sudden end. On September 7, 1855, he was gold panning in the twilight when a stream bank collapsed. He was partially buried and left paralyzed. True to form, he recorded the event in his diary: "Hurt mostly in the small of the back—Spine nearly broken. . . . I was in agony—My lower parts paralysed. . . . No sleep of course—out of my head at times."

For weeks afterward, Doten's diary is the account of his slow, agonizing recovery. Although he would never regain complete mobility, he did improve, but during the process his journal is filled with references to hardships and indignities that included "involuntary stools," "no feeling in my penis or thereabouts," and "no control yet over my evacuations." One month after the accident, Doten summarized his condition in five simple words: "flat broke and broke back."

Three months after the accident, Doten left Gold Country and moved to San Francisco to convalesce and live with his sister. He refused to return to Massachusetts, partly embarrassed that he had not earned a fortune but also hopelessly enamored of freewheeling, liberated Gold Rush California. As he mended, Alfred Doten worked as a farmhand and ranch manager in the Santa Clara Valley, and he put his years of observing and journaling to good use as a newspaper correspondent.

In 1863, Doten followed the silver rush to Nevada. Failing once again to strike it rich, he became a well-respected reporter for several

Nevada newspapers, including the *Virginia City Daily Union* and the *Virginia City Territorial Enterprise*. At the *Daily Union*, he was an acquaintance of a young Samuel Clemens, before he became Mark Twain. In 1872, Alfred purchased the *Gold Hill Daily News* and, under his leadership, the newspaper became one of the most important in the Comstock Lode.

Alfred Doten married and had four children, but his promising journalism career was drowned in a flood of liquor. As his alcoholism worsened, and following some unwise mining investments, the *Gold Hill Daily News* fell into financial difficulty and Doten was forced to sell the newspaper. Eventually, he became a reporter in Carson City, detailing the workings of the Nevada State Legislature for the *San Francisco Chronicle*, although he was best known during this period for bumming drinks from state legislators. His wife, Mary, lived in Reno and sent him an allowance, telling Alfred not to come home. Doten's existence devolved into an endless cycle of drunken reminiscing. His barstool became his office.

But Doten kept writing in his diary, faithfully logging every life event, large and small, for years. On October 2, 1900, he wrote: "My '49 arrival in California anniversary—51 years—*Poorer* than when I arrived, but richer in *humanity* and appreciativeness between man and man than ever—Expended about a dollar celebrating the event with other old '49ers."

By November 1903, Alfred Doten was destitute and alienated from his family, living in a low-rent rooming house, and relying on a few dollars mailed periodically by his wife. His physical condition was deteriorating. He aggravated his old back injury and suffered a serious burn on his leg. And still he kept jotting in his journal. On November 11, 1903, he wrote: "I had nobody to visit me—Had to get along as best I could—not even the Dr calling to see me—Bed 11:30—still continues bad as ever—can't let up."

The next morning, Alfred Doten died in his sleep. He was seventy-four years old.

"GAMBLING ON ONE CARD THE FRUIT OF HIS LABOR FOR THE YEAR"

GAMES OF CHANCE

It took real courage to travel thousands of miles to unknown goldfields and then, once there, to leave your destiny in the hands of dumb luck. You might strike it rich or you might go bust. People were gambling with their lives, at the mercy of what Henry David Thoreau called "the world's raffle." But that wasn't all the argonauts were gambling with.

Games of chance offered surcease from the decidedly unromantic daily experience of the gold miner. There was nothing poetic about being numb from the knees down from standing in an icy stream lifting shovelful after shovelful of gravel into a sluice box. It was not pleasurable to tend to your festering blisters, your aching back, or that hacking cough that had lingered for weeks. There was nothing gentle about sleeping in smelly, mud-caked pants and a fraying flannel shirt that reeked of sweat, bacon, and beans. It was not entrancing to endlessly darn threadbare socks that had more holes than your claim. There was nothing utopian in the constant wariness of your mysterious, dubious neighbors in the diggings. A few minutes at a gambling table was an escape, a momentary adrenaline surge, a welcome but often reckless diversion from humdrum reality. For some, gambling became a most intriguing phenomenon in a curious land far from home. But for a few, it became a deadly obsession.

In 1852, North Carolinian Hinton Rowan Helper found himself in rough-and-tumble Sonora, Tuolumne County. Seeking amusement, he visited one of the gambling halls. Helper watched with fascination the conduct of a Massachusetts gambler named Ned. Helper wrote that Ned "handled the cards with so much grace, skill and agility, and seemed to be so perfectly familiar with every branch of the game, that I could not withhold my admiration." But the situation grew tense when Ned accused a card player of cheating. Angry words were exchanged and numerous threats were lobbed. By Helper's account, Ned, "inflamed with anger, and assuming a menacing attitude, . . . denounced his accuser . . . as 'a pusillanimous liar and scoundrel,' and added, 'G–d d—n you, I'll shoot you!'" The scuffle that ensued quickly escalated into a brawl between Ned, the accuser, and angry bystanders, who were choosing sides in the struggle. A Bowie knife was brandished and, just before it could be plunged into Ned, another fighter produced a revolver and shot the attacker dead. In this "season of dreadful uproar and commotion," Helper continued, "the man who had just committed the homicide was seized by the mob, and, amid loud cries of 'hang him! hang him!' led out to a tree and there summarily executed according to the prompt sentence of the excited multitude." The game resumed. Ned the gambler was nine years old.

Gambling was everywhere. The journals and letters of the Gold Rush are filled with references to the activity. J. Linville Hall, who wrote the first published journal of the era, started it soon after his 1849 arrival in San Francisco. "The principal amusement is gambling," he wrote, "which is carried on to an unlimited extent; and the perfect indifference with which thousands are lost and won would astonish you."

The games were numerous and varied. Among those that would be recognizable to any visitor to twenty-first-century Las Vegas were roulette, keno, poker, and blackjack, which was also called "twenty-one" or "vingt-un." Faro and monte bank, with two-card Mexican and four-card Spanish versions, were also common. But games unfamiliar to us today were just as ubiquitous, including "thimblerig," which was very similar to three-card monte; "lansquenet," a German invention; and the French game "Rouge et Noir" (or "Trente et Quarante"). There were exotic and efficiently cutthroat table games like "chuck-a-luck," which entailed betting on the roll of three dice in a birdcage, or "tub and ball," in which a four-inch ball decorated with one to four quarter-size dots was cast into a velvet-lined tub, and whichever of the cluster of dots ended on top was the winner. A well-liked pastime was the mysterious "strap and pin," whose rules have been lost in time. Other activities were also subjects for betting. There were the infamous bear-and-bull fights, dog-versus-badger battles, billiard contests and, of course, wagering on time-tested, old-fashioned fisticuffs. There was even betting on ten-pin bowling matches. Cheating was endemic.

All were welcome to partake as long as their gold dust glimmered. William Peters, in a letter from New Year's Day of 1851, informed Miss Emily Howland of New York, "Dear Mademoiselle Emilie, . . . The passion that dominates almost everyone is gambling. Every evening till midnight, you can see the Spaniard and the American, the Indian and the Chinese, the tattooed Savage and the little French master crowding a table of Faro or Monte or Rouge et Noir." John David Borthwick wrote in 1852: "Seated round the same table might be seen well-dressed, respectable-looking men, and alongside them, rough miners fresh from the diggings . . . and little urchins . . . ten or twelve years of age, smoking cigars as big as themselves."

Gambling occurred primarily in the gambling halls, which ranged from extravagant "palaces" to quickly fashioned canvas lean-tos sprinkled throughout the goldfields, from the rudest diggings to the most cosmopolitan centers. The larger enterprises frequently provided free liquor to loosen inhibitions and cloud judgment. The lavish gambling halls, or gambling saloons, offered a swirling phantasmagoria of shimmering lights, bawdy paintings of salaciously recumbent nude women, female card dealers wearing only slightly more than their sisters in the paintings, pungent cigar smoke, ear-splitting music, and a disharmonious chorus of cheers and cusswords. They were devilish cathedrals where, as Chilean historian Benjamín Vicuña Mackenna pithily remarked, "gold was the only God worshipped." Chronicler Samuel Colville observed that, in these establishments, "naked and unmasked depravity, daily, nightly, and unblushingly manifested itself." Colville himself found the gambling halls a tempting trap:

> Almost the only comfortable places of resort were the gambling saloons, which were warm and dry, though fetid with the fumes of tobacco, gin, and other liquors, and the poisonous air which has done its duty in turn to a hundred sets of lungs. In such places men needed not drink as a prelude to intoxication. They could absorb it through nostrils and pores of their skin, and, in addition, bands of music helped the excitement and diverted the self-examination and reflection of those who stood within those alluring hells. Few could see the heaps of gold upon the gambling tables and breathe the air, and resist the influences around and before them. Men entered to avoid the rain and get warm, or through curiosity, saw, bet, and were ruined.

Some, such as the pious, teetotaling, and prickly New England–born physician Dr. Israel Lord, were simultaneously titillated and repulsed:

> [The Sacramento hall] is fitted up like a palace. . . . On one side is a counter, 30 feet long, behind which stand three fine looking young men dealing out death [alcohol] in the most inviting vehicles—sweet and sour and bitter and hot and cold and cool and raw and mixed. . . . Oyster and lobster and salad and sauce and fruit and flesh and fish and pies and cakes appear and disappear with so much rapidity and withal so little noise that it seems rather like a dream than a reality—and then this department is served by females. . . . Everything is got up, arranged and conducted with a view to add to the mad excitement of gambling.

Some, probably fueled by overconsumption of liquid courage, were willing to risk "gambling on *one card* the fruit of his labor for the year," in the words of observer William Peters. The mind-altering mixture of sensory overload, excitement, impaired reasoning, lack of sleep, and youthful bravado led to some curious occurrences, such as this one described by Vicente Pérez Rosales in 1849—an incident that seems straight out of a Hollywood movie:

GOLD IN CALIFORNIA.—"EL DORADO," IN SACRAMENTO.—(SEE NEXT PAGE.)

Gold Rush gambling establishments could range from crude to extravagant. Among the most ornate was the El Dorado in Sacramento. "The El Dorado Gambling Saloon, 1852," *Illustrated London News*, June 5, 1852. Courtesy of the California State Library, Sacramento; California History Section.

> Once I had the opportunity to observe a game in which a crafty Oregonian was taking part. He approached the table, and without a word placed on a card a bag that must have contained about a pound of gold dust. He lost. As silently and gravely he placed another bag of the same size, and again lost. Then, with no change of countenance he took from his belt a thin snake [a pouch] that must have held about six pounds of gold. Placing it on a third card, he drew his pistol, cocked it, aimed it at the man who was cutting, and calmly awaited the result. This time he won. "So I win, eh?" he said sarcastically, and with expressionless face picked up his winnings. "I'm in luck tonight." And he disappeared.

A few self-righteous forty-niners tut-tutted and wagged their fingers and claimed that they would never, ever consider a wager, but the reality is that gambling was an inescapable vapor that enveloped the Gold Rush society at its height. As the 1855 *Annals of San Francisco* succinctly concluded, "Every body did so." Writing for more judgmental readers in the future, memoirists sometimes conveniently dismissed their Gold Rush gambling as youthful indiscretion or alcohol-induced moments of insanity. But occasionally a letter will truthfully address the personal impact of betting. As William Peters informed Mademoiselle Emily in 1851, "I have to admit that I was not exempt of that contagion but I have lost enough to feel disgusted and I no longer gamble; bah! It is not my destiny to be rich, so I don't expect it and do not envy those who are, for they don't seem happier than those who aren't."

By 1854, gambling was falling out of favor in California. The influx of disapproving newcomers, many with families, and the public disgust at rampant crime associated with gambling led the 1855 California State Legislature to introduce and approve Senate Bill 149, which banned a few methods of gaming. It was the first step in a thirty-year effort to pass antigambling legislation with teeth—an effort that eventually led to a ban on virtually all forms of gambling in the state. By those who still felt the itch to wager, however, the restriction was simply, and widely, ignored.

"THERE IS NO PERSUASION MORE ESTEEMED FOR MORAL CONDUCT"

JEWS IN THE GOLD RUSH

In 1854, a wagon train headed west to California from Nebraska. Among the company of one hundred wagons and nearly a thousand sojourners were two recent arrivals from Breslau, Germany: Julius Brooks and his seventeen-year-old bride, Isabella, known to all as Fanny. Julius had already spent five years in California, and when he returned to Breslau a year earlier and told Fanny, his niece, a series of fabulous, shamelessly exaggerated stories of the California Gold Rush, she set her heart on reaching the land of gold. His tales were in large part fantastic, but no more so than the American newspaper accounts that made their way to Breslau and seemed to corroborate everything Julius claimed. Uncle Julius planned to return to California, and Fanny begged him to let her accompany him. He agreed, but only if they married.

They set sail from Hamburg on a three-week ocean voyage to the New World. After five months in New York, the couple traveled to Galena, Illinois, a major supply hub and riverboat connection to the wagon train jumping-off points on the Missouri River. Julius and Fanny secured ship passage from Galena to join a caravan departing from Florence, Nebraska, a few miles north of the recently founded city of Omaha. There they bought a covered wagon and two little mules.

Julius and Fanny Brooks began their journey in June 1854. As Fanny later told her daughter Eveline, they faced bitterly cold weather, mammoth thunderstorms, deep snow, intractable mud, and unbearable heat. She encountered giant buffalo herds that sent massive clouds of black dust spiraling skyward. Fanny spotted the bleaching skeletons of thousands of draft animals that had died along the treacherous route and the forlorn graves of travelers who sadly would never reach California. Each time they forded a river, their wagons filled with icy water and frequently drenched their clothes and destroyed their provisions. On the road west, Fanny miscarried her first child. Hers was a remarkable story of perseverance, and similar to those of many other pioneering women on the wagon trains. Except for one thing: Fanny Brooks was the first Jewish woman to cross the plains to California.

Religion was on an extended vacation during the Gold Rush. The peculiar societal vortex of the era rapidly mixed different cultural perspectives, political viewpoints, and religious rituals into a fascinating, roiling philosophical brew that left some perplexed and others profoundly troubled. For most, if religion was considered at all, it was an afterthought. The majority simply put religion aside during their adventure, reckoning that as transitory goldfield denizens they could afford to stable their religious principles for the time being. As one Missouri miner confided to Reverend William Taylor in 1850 San Francisco, "I knew I couldn't carry my religion with me through California, so when I left home in Missouri I hung my religious cloak on my gate-post until I should return."

Despite being partially submerged, religion remained a potent undercurrent in the Gold Rush as a touchstone, beacon, and compass. In a world of dozens of ethnic groups and differing personal values, Jewish people were an integral part of the mix, sharing their practices, literature, and personal experiences. Jewish communities in Europe, North and South America, the Middle East, and even Australia were not immune to the contagion of gold fever, and they rallied to the opportunity to seek the shining bounty. An estimated ten thousand Jews joined the rush to Gold Country, and by the end of the era Jews represented nearly 8 percent of San Francisco's population—the highest percentage in any American city after New York. While many Jews remained in cities, a sizeable contingent headed to the goldfields to seek their fortunes.

In California, there were fewer restrictions and less discrimination than many other places, but the Jewish population still faced limitations, most notably an 1858 statute that mandated the following:

> No person, or persons, shall, on the Christian Sabbath, or Sunday, keep open any store, warehouse, mechanic shop, work-shop, banking-house, manufacturing establishment, or other business house, for business purposes; and no person or persons shall sell, or expose for sale, any goods, wares, or merchandise on the Christian Sabbath, or Sunday.

This "Act for the Better Observance of the Sabbath" denied commercial prospects to Jews who closed their businesses on Saturday, the Jewish Sabbath, while Christian merchants could remain open. The law was blatantly discriminatory, and yet some Jewish merchants supported the proscription in the interests of civic harmony. The law was declared unconstitutional in 1858, but it was reenacted in 1861 and stayed on the books until 1883.

Some Jewish migrants failed in California, and some yielded to the seductive allure of readily available Gold Rush vices such as gambling and alcohol, but, as scholars of American Judaism have repeatedly concluded, the majority of Jews in the Gold Rush found success and were recognized for their critical contributions in establishing prosperous mining towns and commercial hubs. In doing so, they retained their cultural and spiritual identity and developed as an energetic and influential component of the Gold Rush world. Eliza Farnham, a prominent reformer of the era, praised the Jewish community as "much-abused but elastic and persistent."

It is sometimes difficult to find these praiseworthy accounts, however, as a disturbing veil of anti-semitism clouds the historical record. There is no avoiding the reality that Jews were generally described in brutally offensive terms during the period. As the Reverend William Taylor, a twenty-eight-year-old Methodist street evangelist, summed it up, they were the "money-loving, Sabbath-breaking, God-forgetting, Christ-rejecting Jews." John David Borthwick, a Scottish artist and caustic Gold Rush commentator, echoed the prejudices of many when collectively referring to Jews as "unwashed-looking, slobbery, slipshod individuals."

But investigate further, and individual viewpoints were decidedly different. No longer was a Jewish man, woman, or child seen as a stereotypical representative of an entire culture; they were neighbors, colleagues, and friends. Robert Levinson, a leading modern historian of the Jewish experience in the Gold Rush, noted that "one of the most pleasant facts of [Gold Rush] history . . . was the complete absence of ill will directed against [individual] Jews." Levinson and others in academia observed that other ethnic groups experienced the same phenomenon during the Gold Rush, but to a much lesser degree. This lack of outward animosity toward Jews was especially true in the goldfields and mining camps most conspicuously in the primary social and religious centers of Gold Country Jewish life: Nevada and Tuolumne Counties.

Many Jews tried their hands at mining, but few made their livings directly from gold, either as miners or as managers. The few who directly attempted the nitty-gritty of placer mining experienced the same fate as the vast majority of gold seekers, with most of them trying, failing, and then moving on to more lucrative pursuits. Many would have seconded the opinion of Polish Jew Abraham Abrahamsohn, who memorably described his unsuccessful 1851 mining effort near Placerville in this fashion: "Anyone who thinks that roast pigeons are flying around here on golden wings, just waiting to be plucked and eaten, should stay home." A handful were involved in gold-producing enterprises as company officers, and sometimes they became involved when miners transferred their claims to Jewish merchants to retire their debts. And indeed, some stereotypes worked in favor of the Jewish community. Jews had long been said to have a keen business sense, and this often led them to be considered ideal for positions of authority in company hierarchies. Henry Rothschild of Nevada City, for instance, served as the secretary of three separate mining companies. But most Jews were merchants, operating groceries, haberdasheries, general stores, and outlets for hardware and mining equipment. Jewish storekeepers were ubiquitous and important economic leaders, providing daily necessities for the towns and cities springing to life in the foothills. In 1861 Grass Valley, seventeen of nineteen clothing and dry goods stores were operated by Jews, and the five cigar stores were all run by Jewish merchants. Four years later, booming Grass Valley boasted twenty-four apparel shops, twenty of which were run by Jewish retailers. Jewish holidays, such as Yom Kippur, were publicly announced so that city residents would not be surprised by midweek shop closures.

In Sonora, Emanuel Linoberg set up shop, or rather shops, in 1851. Linoberg was a Polish Jewish merchant who owned several establishments in the Southern Mother Lode camp, including a large retail store, an entertainment hall, a gold mine, and, co-run with a medical doctor, the "Russian Steam Bath," a frontier spa and hospital. He also dabbled in transportation as the owner of a mule train that delivered goods to his store. Today, a prominent thoroughfare near the Tuolumne County Courthouse in downtown Sonora is named Linoberg Street.

One of Linoberg's neighbors was Meyer Baer, a German Jew who owned a clothing store. Baer is believed to have been one of the first to sell pants produced by the San Francisco Jewish clothing manufacturer Levi Strauss. Down the street was Michel "Big Mike" Goldwasser, a Polish Jew who owned a rough-and-tumble saloon. In hard-drinking Sonora, the business initially did well and Big Mike was joined by his wife and two children, but by the early 1860s, with the saloon struggling as mining began to diminish, Goldwasser and family left Sonora for other opportunities. They moved first to Los Angeles and then to Arizona, where Goldwasser opened a chain of profitable mercantile stores. Michel Goldwasser's grandson was Barry Goldwater, a United States senator from Arizona and the 1964 Republican candidate for president.

The influence of the Gold Rush's Jewish population was more than just commercial, and frequently Jews were civic leaders and guiding lights of the community. Linoberg served on Sonora's first town council, Jacob Kohlman was elected as a Nevada City town trustee in 1857, and there are many accounts of Jewish citizens holding positions of power with fire companies, Masonic and Oddfellow lodges, political party conventions, and charitable and relief organizations.

Religious observances and charitable activities were also woven into the fabric of daily life. In 1852, Aaron Baruh, a native of Bavaria, settled in Nevada City after two years in New Orleans. He married Rosalie Wolfe and ran a series of stores, including clothing, grocery, and liquor enterprises. Despite his shops being completely destroyed *twice* by fires in 1856 and 1858, Baruh persevered and was a respected resident of Nevada City for over fifty years. The Baruhs were officers in local chapters of Jewish fraternal organizations, such as B'nai B'rith. Often the

Fanny Brooks was the first Jewish woman to cross the Great Plains to California during the Gold Rush. Fanny Brooks and children, c. 1859. University of California, Berkeley; Bancroft Library; BANC PIC 1992.047:2–PIC.

first relief agencies in Gold Rush towns were established by Jews. In late September 1855, Grass Valley Jewish merchants established the Hebrew Benevolent Society and raised money to assist those devastated by a massive fire that had leveled the city a few weeks earlier. Targeted publications highlighted Jewish activities in the goldfields, both spiritual and secular, and Jewish publishers provided news for the general public as well. Henry Meyer Blumenthal was the first owner and publisher of the *Grass Valley Union* newspaper, established in 1864 and still in existence today. These business, social, and publishing networks raised awareness of Judaic culture in the mining camps and also provided a conduit through which Jews beyond California could learn about Gold Rush society.

As Jewish immigrants were appreciated in the larger community, they in turn wholeheartedly welcomed non-Jews to their services and soirees. In an 1852 letter to Isaac Leeser, the influential editor of Philadelphia's *The Occident and American Jewish Advocate*, the first general-interest Jewish periodical published in the United States, Aaron Rosenheim details the mutual regard between the two groups, in evidence during the Yom Kippur service held in Nevada City's Masonic Hall:

> The Masonic fraternity of this place having been made acquainted with our request very generously tendered us the free use of their spacious Hall[;] the room was approbiately [*sic*] furnished and the Ceremonies conducted with dignity and ability, [and] the Room was crowded with Visitors, who were anxious to visit our ceremonies. Among the visitors were the first Citizens of the place, the Judges of Courts, &c, and all expressed their entire satisfaction at our ancient and holy ceremonies and proceedings which were conducted with profound respect.

During the Gold Rush, the coming together of so many different types of people was not always seamless, but in many cases it was a model of tolerance and acceptance.

And what of Fanny Brooks?

Her arduous journey across the plains ended in California in 1854 after an enjoyable respite in Salt Lake City, a developing outpost that

Fanny found delightful. Once in California, Fanny and Julius briefly settled in Marysville, Yuba County, and then opened a store a few miles away in the brand-spanking-new hydraulic mining camp of Timbuctoo, located about fifteen miles west of Grass Valley. Within months of its birth, boomtown Timbuctoo boasted twelve hundred residents, plus churches, stores, hotels, saloons, gambling halls, a sturdy Wells Fargo office, and the ever-growing Brooks family. Their daughter Eveline was born in Timbuctoo, as were two other children. In her memoirs, Eveline recalled, "Father opened a general merchandise store, but the town was small and not a good business center." Today, nothing remains of Timbuctoo except crumbling foundations and fading historical remembrances.

The Brooks family left Timbuctoo as hydraulicking waned. They gravitated first to San Francisco and then to Portland and Boise before ultimately returning to Salt Lake City in 1864 as the first Jewish family in Mormon-dominated Utah. Fanny opened a popular boarding house with a dining room that could accommodate forty patrons. In 1868, Mormon patriarch Brigham Young proclaimed that Mormons were prohibited from doing business with non-Mormons, and although many local establishments collapsed as a result, the feisty Fanny would not submit meekly. She pled her case directly to Young, and he relented, allowing the Brooks family an exemption from the edict. The family business expanded and prospered.

The tale of Jews in the Gold Rush is consequential but largely hidden from our present consciousness. The stories of Fanny Brooks, Aaron Baruh, Emanuel Linoberg, Abraham Abrahamsohn, "Big Mike" Goldwasser, and myriad others are too often forgotten but they resonate as quintessential stories of grit and faith. Many of the successes of the California Gold Rush, many current institutions, many enduring communities would not exist without the involvement and encouragement of the Jewish community. As San Francisco attorney Henry Labatt wrote in May 1861, "No place in America is the Jew so well understood, and so readily appreciated, as in this State; and nowhere does he more deserve the respect and esteem of his fellow citizens. May it always be so."

"EVERYTHING LOOKS FORLORN AND WRETCHED"
STORMS AND FLOODS

When it began raining in November 1849, the rivers started rising and gold seekers had a difficult choice to make. Old-timers warned that the onset of the rainy season would quickly enlarge the Gold Country's streams and that mining the many tributaries where most of the camps were located would be virtually impossible. Water was a crucial tool in gold mining, but miners could never fully control the water. Floods were devastating and deadly, destroying property and shattering dreams.

As the rains intensified in 1849, miners weighed whether to stay in vulnerable floodplains or head to safer ground away from their claims. Most miners chose to weather the storms on more secure terrain, establishing winter quarters on the ridges above Mother Lode canyons and arroyos while they waited for drier and more profitable days in spring and summer. They spent their time repairing their cabins or cutting firewood or simply communing with their fellow argonauts. But not all stayed near the camps. In Sacramento City, observers noted the appearance of miners retreating from the rising waters to the perceived safety of the valley town. Already teeming Sacramento was bursting at the seams and could barely absorb the new arrivals. As forty-niner diarist Dr. Israel Lord wrote, "utter confusion and total disorder" descended on the boomtown under the combination of the harsh weather and an influx of fleeing miners. By mid-November, the relentless rain had turned the dirt streets of the city into quagmires. Jonas Winchester, a recent arrival from New York, recounted in a letter from November 19, 1849, that the residents "are half a leg deep in filth and mud, rendering getting around awful beyond description. . . . The city is one great cesspool of mud, offal, garbage, dead animals, and that worst of nuisances consequent upon the entire absence of outhouses."

The downpours gathered strength and the Sacramento River began to rise perceptibly in early December 1849. But Sacramento residents went about their business, oblivious to the possibility of flooding. As Dr. John Frederick Morse remarked in his 1853 *History of Sacramento*: "The reckless spirit of speculation had declared an inundation as out of the question, if not physically impossible. The very air was tremulous with oft repeated assurances that the town plot had remained free of floods."

And then the rainstorms got worse. It was a terrifying start to the New Year.

On January 1, 1850, Josiah and Sarah Royce and their young daughter Mary reached Sacramento after a long wagon journey from Iowa. They planned on purchasing a lot in Sacramento and opening a grocery and provision business, but, for now, they were content to pitch their tent on the banks of the Sacramento River. When the Royces headed for California in April 1849, Sacramento City had four houses and a few dozen inhabitants; by the time they arrived eight months later, the city boasted ten thousand residents.

On the day the Royces set up camp, the skies darkened ominously as a fierce storm approached. The deluge arrived with a vengeance and, as Sarah recalled, "rain, rain, rain would pour down for hours." When it was over, more than thirteen inches of rain had fallen, and within a day or two the rivers bracketing the city began to swell rapidly. Sacramento City was located on an ancient lakebed at the confluence of two major rivers, the Sacramento and the American. Too much rain could spell disaster.

On January 9, Josiah Royce burst into the family tent and anxiously declared that the riverbanks had been breached and "the water's coming in." About a hundred yards distant on slightly higher ground, a two-story wooden building was under construction. Still unfinished, the structure did have a roof and stairs leading to the second floor, and so Sarah bundled up little Mary, grabbed a few necessities, and raced toward what she later called "our little ark." When they reached the structure, Sarah recalled that she "could hear the rippling and gurgling as the [water] rose higher and began to find its way into crevices and over sills in the lower story." For a few hours they were safe from the flood, but soon it became evident that the waters would inundate the lower story. More than fifty soaked and fearful Sacramentans sought refuge in the building. "Voices of all tones," Sarah remembered, "were heard there, from the stalwart bass to the shrill cry of infancy." They scrambled to the upper floor, hunkered down, and awaited their fate. Within twelve hours, the water in the city rose nearly ten feet. The next day the Royces and others were shuttled to higher ground by rescue boats that docked at the building's upstairs windows.

Josiah and Sarah no longer had any desire to purchase property in Sacramento. After enduring what Sarah called a "tedious week" in the flooded city, the family was able to secure passage aboard the steamboat *McKim* to San Francisco. Once there the family first lived in a hotel, then in a renovated tenement for fifteen months. Seeking better prospects, they eventually headed back to the goldfields. After visiting several mining camps, they settled in Grass Valley, Nevada County, in 1855, where Josiah was a shopkeeper and Sarah occasionally taught school. The Royces lived in Grass Valley for the next twelve years.

The Royces were lucky. The Great Sacramento Flood of 1850 took lives, displaced thousands of residents, destroyed millions of dollars in merchandise, and obliterated sidewalks, streets, and the impermanent tent settlement near the river. Barrels, lumber, canvas remnants, and the bloated carcasses of dead cattle and mules floated on the submerged streets. The flood transformed Sacramento into a city of second-story survivors in the middle of a vast lake.

The smell was oppressive. Raw sewage, putrid animal corpses, rotting produce, decomposing groceries, and decaying vegetation blended with fetid floodwaters to assault the senses. The *Placer Times* reported that

> the entire city, within a mile of the embarcadero, was under water. The damage to merchandise and to buildings and the losses sustained by persons engaged in trade is very great—vast quantities of provisions and goods having been swept away by the rushing waters. The loss in livestock is almost incalculable; many persons have lost from 10 to 50 yoke of cattle each, and horses and mules have been carried down the stream in great numbers.

Teenager Sallie Hester confided to her diary, "Wish I was back in Indiana. . . . Snakes are plenty [here]. They come down the river, crawl under our beds and everywhere."

Dr. John Frederick Morse noted that "when the deluging waters began to rush in and overwhelm the city, there was no adequate means of escape for life and property; and consequently many were drowned, some in their beds, some in their feeble attempts to escape, and many died in consequence of the terrible exposures to which they were necessarily subjected." Patients from a hospital near the riverbank drifted on cots, moaning and weakly appealing for help. Other sick and injured people were abandoned to their own devices, often with horrific consequences. Dr. Morse described one poor victim who, after several days alone and wrapped in a blanket, was found suffering chronic diarrhea, his blanket caked with excrement, and his clothes a strange grayish tint that seemed to be alive. Upon further inspection, it was discovered that "the grey was found to [be] a perfect coating of . . . execrable animals, technically called pediculæ [lice], and of that abominable species that prefer habitation upon the bodies of neglected and filthy individuals."

Dead bodies were fished from the flooded thoroughfares, and the total number of dead and injured was never determined. Many of the deceased were deposited at a lumberyard, which constructed coffins for the multiplying dead. One enterprising citizen, a miner who always carried his gold dust with him, offered his rowboat to transport the departed to higher ground and bury the coffins. With seven pounds of gold dust (worth about $61,000 today) in his jacket pocket, he placed a single coffin crosswise on his tiny boat and headed to the burial ground. Halfway there, his dinghy began to sink. Unwilling to surrender his gold, the extra weight in his jacket pulled him under, and although he valiantly struggled, popping to the surface several times, he finally slipped beneath the water and drowned. The coffin floated to safety.

Yet, surrounded by devastation and doubt, Sacramento residents exhibited a resistance to meek submission. Or, perhaps, they were just in denial. Dr. Morse recalled that as the "sweeping currents . . . were running through the streets . . . no man could have found among the losers of property a single dejected face or despondent spirit. There were no gloomy consultations, no longing looks cast upon the wakes of absconding produce, no animosities excited." In fact, Morse marveled, "the city seemed almost mad with boisterous frolic, with the most irresistible disposition to revel in all the joking, laughing, talking, drinking, swearing, dancing and shouting." At the Eagle Theater on the Sacramento riverfront, a well-attended play carried on as the flood rose. The patrons simply moved to the back of the theater, stood on the chairs, and amused themselves by pushing their friends into the ever deepening water.

As the flood receded, merchants reopened their establishments and renovation commenced within a few days. Few rebuilt with consideration for future flooding. Surveyors were dispatched to establish locations for a legion of levees to ring the defenseless city, and a levee commission was formed, but although the need for levees was obvious, there were no construction funds available. In April, voters approved a special tax assessment for permanent levee construction, and nine miles of new levees started to emerge soon afterward. The flood shields were front-page news, and by October 1850 the *San Francisco Daily Alta California* was referring to Sacramento as "Our Sister, the Levee City." The earthworks lulled hopeful Sacramentans into a sense of false security, and the levees held for more than a year. And then, on March 7, 1852, floodwaters shattered a sluice gate on the Sacramento River, and the city filled as quickly as an unattended bathtub. Within hours, water stood twelve feet deep in some downtown locations. No lives were lost, but the economic impact was severe, despite the early protection the levees provided.

Citizens demanded additional levees, and the city appropriated $50,000 for that purpose. The newest levee, a supplementary bulwark against watery catastrophe, was completed in November 1852. Three weeks later, a forty-foot gash opened in a levee on the American River, and it rapidly expanded to 150 feet in width, inundating Sacramento once again. The *Daily Alta California* reported, "The water was running through Eighth Street, some six feet deep. Several lives were supposed to have been lost. One man was seen floating down the river on the top of his house. At the foot of L Street, a whole block is afloat; the Eagle Saloon is washed away and is floating round." This 1852 flood was two feet higher than the flood of 1850. By New Year's Day 1853, the Sacramento River had risen twenty-two feet, and it went even higher the next day.

Outraged citizens pleaded for improvements, renovation, and increased funding for flood protection. Sacramento city officials responded favorably, and an upgraded levee system safeguarded the susceptible community for several years. But then December 1861 dawned.

Over forty-three days a series of extreme rain- and snowstorms swept across Northern California. The figures are stunning. On December 8, 1861, six inches of rain fell in Nevada City, Nevada County, in twenty-four hours. The same storm pelted Red Dog, a mining camp east of Nevada City, with eleven inches in forty-eight hours. During December, Red Dog witnessed forty-five inches of rain, and over the entire rainy season, the tiny settlement received more than nine feet of rain. Nearby Grass Valley had significant rainfall for sixty-six out of seventy-five days. From November 1861 to the end of January 1862, Sonora, in Tuolumne County, accumulated eight and a half feet of rain. At least fifteen feet of snow was deposited in the Sierra Nevada.

When it was all over, most of the Central Valley, nearly six thousand square miles, was transformed into a colossal inland sea, with depths approaching thirty feet in some places. William Brewer described the widespread devastation in his 1864 book, *Up and Down California*:

> Thousands of farms are entirely under water—cattle starving and drowning. All the roads in the middle of the state are impassable; so all mails are cut off. The telegraph also does not work clear through. In the Sacramento Valley for some distance the tops of the poles are under water. The entire valley was a lake extending from the mountains on one side to the coast range hills on the other. Steamers ran back over the ranches fourteen miles from the river, carrying stock, etc., to the hills. Nearly every house and farm over this immense region is gone. America has never before seen such desolation by flood as this has been, and seldom has the Old World seen the like.

One-quarter of California's cattle drowned. In the valley flood zones, one house in eight was destroyed, and virtually all the remaining structures suffered some damage. Thousands of residents died.

Throughout Northern California, the floodwaters reached alarming depths. North of San Francisco, the town of Napa was under four feet of water, and Rio Vista in the Sacramento Delta was flooded six feet deep. The Gold Country communities of Knights Ferry (Stanislaus County) and Mokelumne Hill (Calaveras County) saw nearly every building ripped to shreds by avalanches of mud. A landslide in Volcano, Amador County, roared through the village, killing seven.

Warm spring rains melted the snowpack, and the resultant flood washed tons of accumulated hydraulic mine debris into the watersheds. This murky slop, a witch's brew of decomposed granite, sediment, rotting vegetation, clay, and mercury called "slickens," hurtled downstream with devastating effect. The sludge topped the riverbanks and fanned out over the Sacramento Valley floodplains. Fertile farmlands were buried under two to seven feet of the sediment. The levees surrounding Marysville, Yuba County, burst and allowed six feet of this opaque gunk into the city.

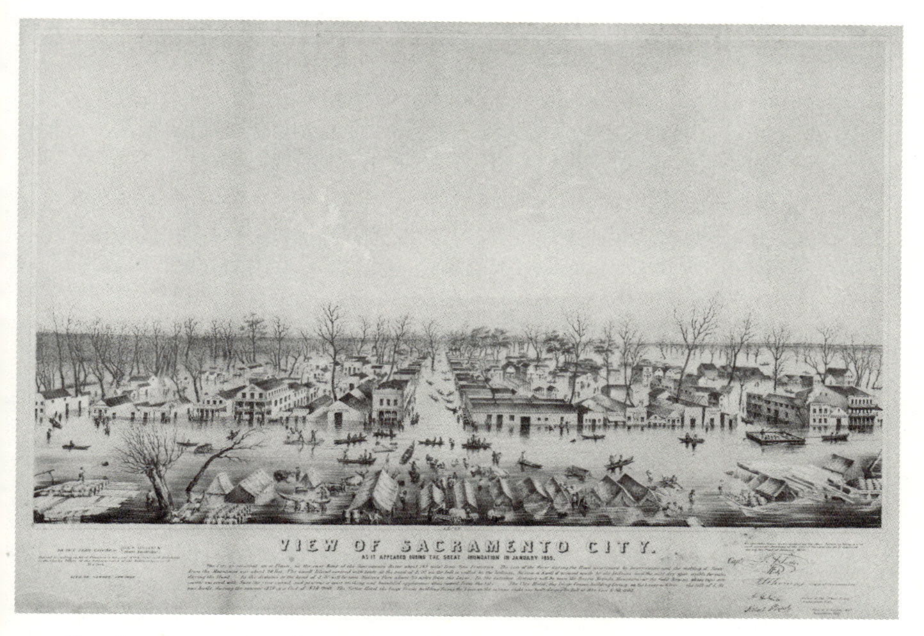

Flooding was common during the Gold Rush, and in 1850, Sacramento suffered a flood so massive it made the history books. "View of Sacramento City as It Appeared during the Great Inundation of January 1850," lithograph by Sarony, c. 1850. Courtesy of the California State Library, Sacramento; California History Section.

Northern California suffered the brunt of the storms, but Southern California was not unscathed. Los Angeles received sixty-six inches of rain that season—more than four times the average rainfall. Mucky floodwaters accumulated everywhere in the southern portion of the state, even forming substantial lakes in the Mojave Desert.

In Sacramento, the revamped levee system was no match for these torrents of Biblical proportions. Water relentlessly raced into every low-lying nook and cranny. River sediment and hydraulic mining residue left ridges of sand and slop eight feet deep in the city center. William Brewer memorably illustrated the continuing aftermath:

> Such a desolate scene I hope to never see again. Most of the city is still under water. . . . No description that I can write

> will give you any adequate conception of the discomfort and wretchedness this must give rise to. I took a boat and two boys, and we rowed about for an hour or two. Houses, stores, stables, everything, were surrounded by water. Yards were ponds enclosed by dilapidated, muddy, slimy fences; household furniture, chairs, tables, sofas, the fragments of houses, were floating in the muddy waters. . . . The basements of the better class of houses were half full of water, and through the windows, one could see chairs, tables, bedsteads, etc., afloat. . . . Over most of the city boats are still the only way of getting around. . . . Not a road leading from the city is passable, business is at a dead standstill, everything looks forlorn and wretched. Many houses have partially toppled over; some have been carried from their foundations, several streets (now avenues of water) are blocked up with houses that have floated in them, dead animals lie about here and there—a dreadful picture. I don't think the city will ever rise from the shock, I don't see how it can.

On January 10, 1862, California's newly elected governor, Leland Stanford, was to be inaugurated. Floodwaters were rising one foot per hour on that day and the capitol building was swamped. The office of the state treasurer was submerged under three feet of water. But the ceremony proceeded despite the waterlogged mess. Stanford arrived at the ritual by rowboat and, following a hasty administration of the oath, departed by boat to his flooded downtown mansion a few blocks from the capitol. Stanford was obliged to enter his home by rowing to and then scrambling through a window on the second story. The California State Legislature despaired. Unable to adequately perform their duties in the swampy, stinking mire, the legislative body officially retired to San Francisco in late January 1862 while Sacramento dried out.

After the destructive storms and floods of 1861–62, Sacramento developed an extensive flood-prevention plan. The city repaired and expanded its levees and began a seven-year project to raise the downtown street level ten to fifteen feet. Governor Leland Stanford added a third story to his mansion and left the ground floor empty as a way of dodging future flood damage.

"A WIND TURNED DARK WITH BURNING"

THE PLAGUE OF FIRE

In 1852, Sacramento was a bustling city of thirteen thousand residents, but it was vulnerable to devastating fires. The community was dotted with the occasional "fireproof" brick buildings, but most structures were hastily constructed from wood, as one would expect in a boomtown. At 11:10 P.M. on November 2 a fire broke out in the Madame Lanos and Company Dry Goods and Fancy Store on J Street. Several blocks away on K Street, shopkeeper Collis P. Huntington, who later cofounded the Central Pacific Railroad and built the western half of the first transcontinental railroad, was busily fighting the flames in his hardware store with his wife, Elizabeth, and their female servant, whose bonnet caught fire as they struggled with wet blankets and sacks to save the store. It was to no avail. As Huntington later wrote to his younger brother Solon, he kept fighting "until the Back Counter wos all on fire and I saw that no effort of mine could Save it and then I wraped a wet Blanket around me and . . . left the Store through a sheet of flames and Saw the earnings of years consumed eaven to the last dollar." His eyes were seared and he had difficulty seeing for several days. Collis Huntington was fortunate to escape alive. Elizabeth Huntington recalled that "a Mr. Brigam who was a very smart young man was burned to death on the same Block."

A few would later claim it was arson committed by disgruntled voters disappointed by the results of the election held earlier that day, but most believed it started with an accident at the dry goods shop of Madame Lanos. As the *Sacramento Union* reported on November 4, the fire, whipped by stiff breezes, spread rapidly, baring its "lurid fangs" and "enveloping [the city] in a sheet of fire." Within three hours, 90 percent of Sacramento was reduced to smoking ruins. The *Union* described the scene:

> Men gathered in crowds—some looking on the waves of fire as they rolled from house to house, and street to street, bewildered, fascinated, and apparently paralized [*sic*] at the terrible work of destruction going on before them—while others were hastening to save their goods and effects, before the fire should reach them.

Draymen wheeled their low, strong carts to the cusp of the firestorm and offered to move business and personal belongings to safety—for a fee, of course. The rate started at $50 per load.

There was a dreadful accounting when the fire died out. At least six lives were lost, more than twenty-five hundred buildings were destroyed, and property losses were conservatively estimated in excess of $10 million ($305 million in today's money). But Sacramento began rebuilding even as the mountains of debris still smoldered. Wooden buildings were dismantled in San Francisco and shipped to Sacramento for reassembly. Huntington and other merchants took out loans and began reconstruction. Within days, new stores, hotels, liveries, warehouses, and residences sprang to life, "before the old ones were fairly done burning," as the *Union* noted. "Others too, . . . contracted to have buildings put up on the lots occupied by those destroyed. . . . All have gone to work with a determination to do what man can do to restore the prosperity, beauty, and business of the city." Six weeks after the fire, on December 13, the *Union* reported that "the city to-day exhibited a most bustling and active [commercial] appearance. Large quantities of goods changed hands, and have been dispatched to the Interior. Buyers evidently purchased more freely and in greater quantities than they have done for some times past." Sacramento was back in business. One year later, the November 5, 1853, edition of the *Sacramento Union* carried a prominent advertisement for the Madame Lanos and Company Dry Goods and Fancy Store, highlighting a recent shipment "Direct from the Eastern market" of "a choice selection of FALL BONNETS of the latest styles, SILK AND STRAW—a variety of the most fashionable description of Autumn Dresses, Cachmere, Delaine, French Merino & c."

From the era's genesis, fire was the plague of Gold Rush California, a recurrent symbol of ruin and resurrection, and Sacramento was by no means the only major California community to be devoured by fire during this time period. San Francisco was repeatedly afflicted, with conflagrations consuming portions of the city in December 1849, May 1850, June 1850, September 1850, December 1850, May 1851, and June 1851. As the flames were once more licking the hills on May 4, 1851, the *San Francisco Daily Alta California* moaned,

> San Francisco is again in ashes. The smoke and flames are descending from several squares of our city; as if the God of Destruction had seated himself in our midst, and was gorging himself and all his ministers of devastation upon the ruin of our doomed city and its people.

Gold Country was not immune from the scourge of fire; virtually every mining camp and growing city was touched by the "God of Destruction." Even the threat of fire in these ephemeral communities could set pulses racing, and the panic was not unwarranted. Maryland argonaut Luther Schaeffer recalled the night of June 14, 1851, in Grass Valley, Nevada County, when the residence of an acquaintance caught fire:

> Fire, fire, fire! the alarm was sounded—every inhabitant felt excited—run, men, run! All the houses were constructed of poles and shingles, lined with muslin or paper, and should the fire spread, good-bye to our village; we had no fire engines, and no water convenient. Fire! fire! Where? Down 'Mill street.' I saw the flames issuing from one of friend S——'s houses; fortunately, it stood alone, and the evening was calm. The building was consumed. The fire extended no farther, and the danger was over.

Fire was a frequent bane during the Gold Rush. Many cities and gold camps were razed, some more than once. "The Great Conflagration at Sacramento, California, 1852." Courtesy of the California State Library, Sacramento; California History Section.

If one were to design a town most susceptible to fire, one might wind up with something akin to the settlements that sprung up across Gold Country. Their temporary structures of wood and canvas made the perfect kindling. Take Sonora, the county seat of Tuolumne County. In 1852 it was a town of several thousand inhabitants, and it exhibited many of the characteristics of a typical Gold Rush camp, albeit with a more eclectic architectural mix. There were impromptu, impermanent canvas shanties nestled next to solid one-story adobe edifices set cheek by jowl with recently constructed wooden buildings. Hotels, fandango halls, the obligatory gambling halls and saloons, stores, and dwellings lined the single main street and dotted the hillsides. Sonora had been founded in the early days of the rush by Mexican miners, commonly called "Sonorans," and some of the buildings displayed what observers described as a distinctive Mexican style: wooden structures, sometimes incorporating adobe, with balconies, high over-arching roofs, and staircases on the outsides of the buildings. These structures were ornamented with filigree and often painted buff or pale blue. Popping up everywhere were also the more recent American-style structures: two-story white rectangular facades fronting much lower clapboard stores or canvas tents. These were strictly utilitarian affairs, with the storefront elevations designed to hold large signs advertising their wares. This architectural assortment of Mexican, American merchant, and temporary diggings-style structures was more pronounced in Sonora than in many other mining camps, but all these structures had one shared trait: they were firetraps.

On June 18, 1852, a massive fire obliterated Sonora in a matter of minutes. Itinerant Scottish author and illustrator John David Borthwick memorably described the frenzied reaction to the conflagration:

> It was about one o'clock in the morning when the fire broke out. I happened to be awake at the time, and at the first alarm I jumped up, and, looking out of my window, I saw a house a short distance up the street on the other side completely enveloped in flames. The street was lighted up as bright as day, and was already alive with people hurriedly removing whatever articles they could from their houses before the fire seized upon them.
>
> I ran down stairs to lend a hand to clear the house, and in the bar-room I found the landlady, *en deshabille*, walking frantically up and down, and putting her hand to her head as though she meant to tear all her hair out by the roots. She had sense enough left, however, not to do so. A waiter was there also, with just as little of his wits about him; he was chattering fiercely, sacréing [swearing] very freely, and knocking the chairs and tables about in a wild manner, but not making a direct attempt to save anything. It was

> ridiculous to see them throwing away so much bodily exertion for nothing, when there was so much to be done.

Borthwick also recalled the fire's aftermath, describing the sad tableau of the dispossessed populating Sonora's hillsides:

> On the hills, between which lay the town, were crowds of the unfortunate inhabitants, many of whom were but half dressed, and had barely escaped with their lives. One man told me he had been obliged to run for it, and had not even time to take his gold watch from under his pillow.
>
> Those whose houses were so far distant from the origin of the fire as to enable them to do so, had carried out all their movable property, and were sitting among heaps of goods and furniture, confusedly thrown together, watching grimly the destruction of their houses.

Within hours, however, Sonora was resonant with the buzzing, energized orchestra of construction. "Before one could look around," Borthwick wrote, "crowds of workers on the long blackened tract of ground which had been the street" were sawing, hammering, digging, sweating, cussing, and erecting frameworks for buildings. As the ground cooled, merchants and residents returned to the smoky town center and fabricated lean-tos and canvas tents, and moved their surviving belongings into these makeshift shops and dwellings. Some simply set up business by draping canvas tarps over piles of merchandise boxes, barrels, and casks.

Often in these razed settlements the first establishment to reopen was the gambling saloon. Sonora was not an exception. The morning after the fire, Borthwick noted that the new saloon was in full operation "but an hour or two after the town had been burned down."

> The same gamblers were sitting at the same tables, dealing monte and faro to crowds of betters; the piano and violin, which had been interrupted by the fire, were now enlivening the people in their distress; and the bar-keeper was as composedly as ever mixing cocktails for the thirsty throats of the million.

John David Borthwick also recorded the extraordinary, indefatigable spirit of the populace to restore a semblance of continuity after fire ravaged the town. This panache, this inability to concede defeat even when faced with devilish infernos, was a repeated theme throughout Gold Rush history. The many who had sacrificed much and traveled far to seek their fortunes in the ice-cold streams and darkened recesses of Gold Country were not so easily deterred. They fought back. They made the phoenix rise.

Perhaps the final word on the resilience of the gold seekers was provided by French argonaut Ernest de Massey, who had repeatedly witnessed the destructive power of fire in Gold Rush–era San Francisco and the subsequent buoyant resurgence of the city. In a letter to his cousin in France, de Massey offered this admiring coda: "One calamity more or less seems to make no difference to these Californians."

"THE BLUE VAULT OF HEAVEN"

ALONZO DELANO AND THE GREAT GRASS VALLEY FIRE OF 1855

On July 11, 1854, a blaze sparked on a bone-dry cabin roof on Neal Street in Grass Valley. Within minutes, an adjoining stable and three houses were fully engaged. No injuries were reported and volunteers doused the flames, but a concerned citizenry called for a permanent firefighting force. Funds were raised for equipment. Members were recruited. History does not remember if the company was adequately trained for their task, but residents fondly recalled October 31, 1854, when the firemen proudly paraded through town in their shiny new uniforms.

Grass Valley grew rapidly as it began to exploit its rich quartz gold veins, and by 1855 it was a prosperous community of thirty-five hundred. Hundreds of wooden structures dotted the city, with only four "fireproof" brick buildings scattered on the landscape. While it was not the rude, wild mining camp it had been a few years earlier, elements of the rough-and-tumble persisted; it was a hard-drinking, hard-working, two-fisted town that was just beginning to develop the trappings of civilization. Saloons were prevalent, gambling was a favored pastime, and women were uncommon. According to an 1852 account, Grass Valley had a grand total of two women in town, and although by 1855 the numbers had increased, women were still relatively scarce.

On March 5, 1855, the town incorporated, and a week later the community elected city officers, including a five-member board of trustees, a marshal, an assessor, a city clerk, and a treasurer. Best remembered among these first officials was the city treasurer, Alonzo Delano, a prominent merchant and celebrated Gold Rush author.

While the new government was destined to be short-lived (the California Supreme Court declared the city's incorporation legally invalid in 1856), the new board of trustees speedily passed sixteen municipal ordinances, and the town marshal just as quickly administered them. On his first day on the job, Marshal John Little made four arrests—one for fighting, one for "fast riding," and two for women wearing male attire in public.

The Grass Valley board of trustees also addressed the issue of fire, a plague that impacted most Gold Rush–era communities. Many towns up and down the Mother Lode had already suffered significant losses due to fire, and Grass Valley itself had experienced many, albeit mostly small-scale, fires and had managed to provide only inadequate response. Hoping to avert a potential disaster, the board passed an ordinance requiring every occupant of every building to construct a cistern holding at least fifty gallons of water and to have four fire buckets for each story. Unfortunately, this ordinance was never enforced. Delay is the assassin of good intentions, and in Grass Valley it led to devastation.

At eleven o'clock on the night of September 13, 1855, a fire broke out in the United States Hotel on Main Street and quickly spread through the heart of Grass Valley. Cries of "Fire! Fire!" would crescendo through streets, suddenly alive with desperate action. Residents battling the firestorm swarmed like ants. But to no avail. The 1880 *History of Nevada County*, published by Thompson and West, described the ensuing chaos:

> All was confusion; the flames were crackling and roaring, licking up the tinder dry buildings in their pathway, and all the undirected efforts of the excited people were futile to stay their onward march. Buildings were pulled down, buckets of water by the hundreds were thrown upon the burning

> houses, wet blankets and other devices were resorted to, but to no avail, for the frame buildings, dried in the long summer sun, burned too fiercely for the flames to be subdued. All night they fought with tireless energy, and never ceased the struggle until the flames expired for want of food to live upon.

The destruction was enormous. More than three hundred buildings were destroyed in an area covering thirty acres. In the downtown business district only the four brick structures escaped. Every hotel and boarding house was ruined. Homeowners and merchants believed their life savings and essential documents, mostly stored in the Wells Fargo vault, were lost to the fire. Dreams were reduced to ashes, and all seemed hopeless. Newspaper accounts, which likely lowballed their estimates, indicated that the loss of property was valued at $400,000—$10 million in today's money.

It is at this point that Alonzo Delano enters to save the day and bolster the town's resolve.

When Alonzo Delano took up residence in Grass Valley in the early 1850s, the mining camp was still bucolic but growing rapidly. Grass Valley, c. 1852, lithograph by R. E. Ogilby. Courtesy of the California State Library, Sacramento; California History Section.

When gold was discovered in California in 1848, Alonzo Delano of Ottawa, Illinois, was ill. While others were catching Gold Fever, Delano just had a regular fever. Consulting his doctors, Delano was given a very curious prescription: the best remedy, the physicians claimed, was to travel to California on an ox-driven wagon.

With that odd medical order in hand, Delano lit out for California in the springtime of 1849 as one of the many emigrant vagabonds headed west. His account of the journey was published in 1854 as *Life on the Plains and Among the Diggings*.

Alonzo Delano was not a miner. Instead he mined the miners as a merchant in various Mother Lode boomtowns. Mostly he failed, and over time he gravitated to San Francisco, where he established a business in 1850. There he prospered, and he eventually moved to Grass Valley, the golden heart of the northern mines. In Grass Valley he set up additional shops, became a banker, and served as the Wells Fargo agent.

And, in 1855, he became vital to the town's recovery after the fire. One of the few items to survive the September blaze was the Wells Fargo vault. As the town still smoldered, Alonzo Delano made a dramatic appearance, as Ezra Dane recalled in 1934:

> Something was moving down the hill from the west end of town. It was a frame shanty, on rollers. And who was the figure in the rumpled frock coat directing its progress? A profile view identified him as Old Block, setting an example of California courage for the citizens. A willing crowd gathered to assist in backing the building up against a brick vault, which was hot but still standing among the ruins where the express agency had been. A few minutes later a ten-foot scantling was nailed over the door, roughly lettered "Wells, Fargo & Co.'s Express Office"—and Old Block, so the county history [of Thompson and West] tells us, "stood smiling behind his counter, amid the smouldering ruins and with the ground still warm beneath his feet, ready, as he said, 'to attend to business.'"

The hot-to-the-touch, debris-covered vault was unsealed, and, to everyone's relief, the important documents and currency inside were shown to have been untouched by the flames. This tiny shed served as the Wells Fargo temporary office and a touchstone of civic renewal.

Before he was city treasurer or town hero, Delano was a writer of note, and his sketches of gold camp life rivaled those of Bret Harte and Mark Twain in popularity. Under his pseudonym "Old Block," Delano wrote mostly humorous vignettes for a wide spectrum of the era's publications, including the *Sacramento Union*, the *California Farmer*, the *Golden Era*, *Hutchings' California Magazine*, and even the *New York Times*. His popular books included *Pen-Knife Sketches, or Chips of the Old Block* (1853) and *Old Block's Sketch Book* (1856).

Marguerite Wilbur, a collector of his works, wrote of Alonzo Delano: "Grass Valley knew him best of all . . . not only as a writer but as a sober and sound citizen who worked for the good of the community. His wiry figure of medium height, lean and erect, with its keen eyes and enormous hooked nose was a familiar sight on the streets." The events of September 13, 1855, would reinforce Delano's reputation as a "sober and sound citizen."

Among Delano's writings is a description published in *Old Block's Sketch Book* of the Great Grass Valley Fire of 1855. He wrote glowingly of the calmness and determination of the townsfolk: "On the eventful night which laid our town in ruins, which left us no cover for our heads but the blue vault of Heaven; . . . did you hear one word of wailing—one single note of despair? No, not one." It is telling of his character that he modestly failed to inform his readers that he had been instrumental in the town's recovery. But Grass Valley citizens would never forget his contributions. Delano was so admired for lifting the town's spirit in their hour of darkness that upon his death in 1874 all the businesses, including the mines, in Grass Valley suspended operations in tribute, and hundreds attended his funeral.

Following the Great Fire of 1855, Grass Valley quickly rebuilt. Before the end of that year, new businesses and homes dotted the blackened hills, and the streets were carpeted with wooden planks. The city government's response to the fiery disaster? Delay. Grass Valley did not organize a regular fire department for three more years.

A CRUMBLING KINGDOM

THE COLLAPSE OF SUTTER'S FORT

John Sutter's story is an oft-told tale of escape, reinvention, caddish behavior, abandonment, and loss. His world was built on personal charm, self-aggrandizement, broken promises, and an economic house of cards fashioned from delayed payments and massive debt. And in California, he found himself at the center of a story that spread like wildfire around the globe: the discovery of gold, on his land.

He was born Johan Augustus Suter in February 1803 in Kandern, in the Bavarian region of what is now Germany. This small Black Forest community was a few miles from the Swiss border, and Sutter considered himself a citizen of Switzerland—a matter of wishing, not reality. He married and had children, but his relationship was tempestuous and he was a distant, distracted father. He strived to rise above his humble origins and gain prominence and wealth, but, time and time again, Sutter's proclivities for luxury, a regal lifestyle, and an inability to say no led him to serious, often dire, straits.

By 1834, his liabilities far exceeded his assets, and the police were looking for him. Sutter chose to secretly leave Europe for America, disappearing in the middle of the night without bidding adieu to his wife or children.

Arriving in New York some months later, Sutter traveled across the United States to Santa Fe, New Mexico, and entered into trading. His business took him to Oregon, Hawaii, and Alaska. In Hawaii, the amiable Sutter struck up friendships with high-ranking Hawaiian officials, and upon his departure from the islands, these leaders presented Sutter with eight Polynesian servants (most likely slaves) called Kanakas.

In the United States, Sutter began the reinvention process that would transform the struggling Swiss merchant into European nobility. At first he claimed he was a captain in the Swiss army, later that he was an officer in the court of Charles X of France, and subsequently that he was the prominent son of a clergyman; he insisted on being addressed by the titles that came with these invented ranks. He also widely proclaimed that it was his destiny to establish a colony in California and to be its potentate.

In 1839, Sutter was granted an audience with the Mexican governor of California, Juan Bautista Alvarado, and Sutter told him of his vision of an inland empire. This vision was at odds with the reality of California at that time: a sleepy province of Mexico with a small population and burdened by tension with the Native population and the threat of invasion by the Russians. The slowly disintegrating missions were the primary commercial network, and the economic potential of the region was considered minimal. It was the stagnant outpost of a troubled realm. Alvarado was impressed not only by Sutter's persuasiveness but also for practical reasons: Alvarado saw in Sutter an energetic man who could make the considerable effort to establish a colony without the costly help of the Mexican government.

Alvarado gave his blessing to Sutter's enterprise. Additionally, the governor granted Sutter a fifty-thousand-acre tract in the Central Valley, which Sutter accepted sight unseen. He became a Mexican citizen and a semiofficial representative of the Mexican government. But his dreams were of personal glory and feudal power.

In August 1839, Sutter began the journey to his new home in the Sacramento Valley. He sailed through the interior river maze searching for a stronghold for his self-styled empire of New Helvetia, named after his not-quite-homeland of Switzerland, also called Helvetia. On board his boat, the *Isabel,* was the initial party of settlers. There was Sutter, three German carpenters, two mechanics, the ship's captain and several sailors, and Sutter's eight Kanakas. After several days sailing, Sutter, theatrical to a fault, announced his arrival in the valley with a bang—by firing a cannon. Eyewitnesses noted that the blast startled flocks of waterfowl and started a small stampede of tule elk.

Following his arrival at the place that would become Sacramento City, Sutter spent several days charting the land before selecting a knoll situated well back from both the Sacramento and American Rivers. Construction began immediately. The first structures were unimposing but functional. William Wiggins, a twenty-three-year-old visitor to New Helvetia in 1840, described John Sutter's residence as merely a "one-story adobe with three rooms" where he "kept upon his table a brace of pistols, and he would allow the wild Indians to come into his room by the scores, oftentimes being . . . entirely alone." The pistols were Sutter's instruments of enforcement, used to deter any manner of offense, from failure to acknowledge Sutter with deference to poor workmanship to stealing to illegal trapping, hunting, and fishing and more serious, violent transgressions. As Wiggins noted, it was widely known that any offenses "would be punished by death": Sutter was judge, jury, and executioner.

Over the next several years, an impressively self-sufficient community emerged from this modest beginning. There was a blacksmith shop, a carpenter shop, and spaces for weavers, coopers, saddlers, shoemakers, and other artisans. Sutter had walls constructed around the settlement, a plastered and whitewashed adobe barricade nearly eighteen feet high and more than two feet thick, encompassing an area estimated to be about seventy-five-thousand square feet. At each end of the enclosure were two cannon emplacements in towers, and a two-story adobe command post, covered with gleaming white paint, was centrally located within the compound. The fort provided security for the first inhabitants and a frontier "castle" for Sutter, imperious master of New Helvetia.

Sutter's frontier outpost grew into his vision—a personal dominion. It featured irrigated fields, orchards, vineyards, sheep, cattle, horses, hogs, and other crops, including a perpetual harvest of enormous debt. It was established as a business and trade center and as a rendezvous point for those who had journeyed over the treacherous Sierra Nevada. Captain Sutter was a trapper, farmer, stockman, merchant, and military ruler. He organized a private army of two hundred men composed of Indians and Europeans. They wore red-trimmed blue-and-green uniforms and they drilled accompanied by fife and drum. Sutter also became known for his brutal control and suppression of the local Indians. As he recalled in his 1876 "Personal Reminiscences," Sutter characterized himself as "everything[:] . . . patriarch, priest, father & judge. . . . I had the power of life and death over both the Indians and white people."

By 1849, Sutter's dominance in the Sacramento Valley was manifest, and Sutter's Fort was a vibrant symbol of that influence. In that first full year of the Gold Rush, Samuel Upham, a forty-niner from Philadelphia, described his arrival and first impressions:

> I arrived in the evening at the *Embarcadero*, or port of New Helvetia, the schooner coming to anchor in a fleet of smaller craft. Here were a few huts situated upon elevated ground, nestling beneath the protection of lofty sycamore and oaks. The settlement of Captain Sutter is nearly three miles from the landing-place, the road leading over a beautiful country, constantly rising as you leave the Sacramento.
>
> Captain Sutter's establishment has more the appearance of a fort than a farming establishment. It is protected by a wall, ten feet high, made of *adobes*, or sun-dried bricks, having a turret with embrasures and loop-holes for fire-arms. Twenty-four pieces of cannon, of different sizes, can be brought to defend the walls. . . . At the gate-way is always stationed a servant, armed as a sentinel. I arrived at the establishment early in the morning, just as the people were being assembled for labor by the discordant notes of a Mexican drum. I found Captain Sutter busily employed in distributing orders for the day. He received me with great hospitality, and made me feel on the instant perfectly at home under his roof.

But the humming beehive of activity and "great hospitality" of Sutter's Fort would not last. As the surge of the Gold Rush pummeled

New Helvetia, Sutter's Mexican land grant came under question, as did those of all grantees in Alta California. As the legal system transferred from Mexican authority to American control, Sutter was unable to legally prove his title to the satisfaction of the American authorities. Beginning in 1849, his lands came under assault from the arriving argonauts, who swarmed his property. Sutter's territory was stripped from his hands, and he was thunderstruck and emotionally paralyzed. Overwhelmed, the Master of New Helvetia watched helplessly as hundreds camped freely around his fort and dismantled his fences for firewood. The millstone of his flourmill was stolen, and looters made off with two hundred barrels of packaged salmon from his storehouse. Sutter pinned his hopes on government restitution and acknowledgement of his land title, but it never came. Meanwhile, his workforce

By 1890, the once impressive Sutter's Fort in Sacramento had nearly disappeared. Only the original two-story main building remained, and it was in a state of advanced decay. "Sketching Class," photograph showing Sutter's Fort by William Jackson, c. 1890. Courtesy of the California State Library, Sacramento; California History Section.

abandoned him for the goldfields, and without them his farms were raided, timber cut, animals stolen. One Gold Rush butcher is reported to have made $60,000 in three months from the sale of Sutter's stolen cattle—an enormous sum even with the inflated prices of the early Gold Rush. John Sutter stopped paying his bills and drank excessively. He wrote in his diary:

> One thing is certain that the people looked on my property as their own, and in the Winter of 1849 to 1850, a great Number of horses has been stolen from me, whole Manadas of Mares driven away and taken to Oregon etc. Nearly my whole Stock of Cattle has been Killed, several thousands, and left me a very small Quantity. The same has been done with my large stock of Hogs, which was running like ever under nobodies care and so it was easy to steal them. I had not an Idea that people could be so mean, and that they would do a Wholesale business in Stealing.

Sutter complained, but he was ignored. He eventually abandoned the fort and moved to other properties in his shrinking domain. In an 1857 article for *Hutchings' Illustrated California Magazine*, Sutter sorrowfully recalled, "By this sudden discovery of the gold, all my great plans were destroyed. Had I succeeded for a few years before the gold was discovered, I would have been the richest citizen on the Pacific shore; but it had to be different. Instead of being rich, I am ruined."

By early 1851, Sutter's Fort was a shell. To provide building materials for the booming Sacramento City, only two miles to the west, the fort was plundered for lumber and shingles. In March 1851, William Prince visited the site and reported: "The mud walls are crumbling & in 2 or 3 years will be pretty much demolished—One family occupy a small part of Sutter's old mansion within the fort—& outside two or three buildings have been erected for a hospital." In December 1851, George McKinstry sadly described the fort's condition to its former commander Edward Kern: "The old fort is fast going to decay; the last time I was there I rode through, and there was not a living thing to be seen. What a fall is there, my fellow!"

All that remained was the two-story central building, now abandoned and forlorn. As time passed, the white-painted walls flaked and crumbled. Sutter's once imposing headquarters was as blotchy as a mangy dog.

For decades, this last vestige of Sutter's Fort was a deteriorating symbol of the changes wrought by the Gold Rush, both for Sutter and old California society. Then, in 1891, the property was purchased by the Native Sons of the Golden West with the goal of rebuilding the fort to approximate its appearance at its height in the mid-1840s. In 1893, the restoration was complete. The revived Sutter's Fort, back from the dead, was donated to the State of California and officially became a part of the California State Parks system in 1947.

THE DEVIL'S CHAOS

HYDRAULIC MINING

It was called "the slickens," and it just kept coming. For more than thirty years, year after year, that yellowish ooze came down from somewhere up in the mountains. It was relentless. Hundreds of acres of prime farmland were slowly covered with a deathly veil of muck. Orchards died, houses were engulfed, flooding was common. The rivers were full of goo, and boats could no longer navigate through the debris. Levees towered over nearby towns. It was a juggernaut of slime, sand, and sorrow.

"Slickens" was the name for the residue created by California's hydraulic mining operations, which began in earnest in 1853. At first blush the technique appeared to be a golden dream, and it was widely employed throughout Gold Country during the Gold Rush and afterward as a way to make laborious gold mining a little more efficient. But it was efficiently brutal: using powerful jets of water to rip away the "country rock" and dirt to get to the underlying gold, entire hillsides were swept away in the search for a few ounces of color. Tons of debris were flushed into nearby streams and canyons, choking the waterways that led to the agricultural centers of the Sacramento Valley. In contrast to the mining being done with pans and long toms, this highly effective technique was not the domain of the solitary miner but required more labor, more investment, and the development of associated infrastructure, such as the construction of extensive dam, ditch, and flume systems, ironworks, and the growth of mining camps devoted to hydraulicking. Hydraulic mining became fully industrialized within a few years, changing the nature of what it meant to be a gold seeker in California.

Before, much of California mining involved working the surfaces of ancient gravel beds or digging shafts or tunnels that followed veins underground—work that was done either alone or in small groups. But these methods left untouched the huge lodes locked in the strata of gravel on the sides of canyons or on land hundreds of feet above riverbeds or even miles from the nearest water source. To access these reserves, miners employed the ancient technique of "ground sluicing," which involved first conducting water to the top of a mine location via flumes and ditches, then letting the stream cascade over the rim to loosen debris, which then fell and could be mined.

Hydraulic mining was a modern variation on ground sluicing in which water was delivered not by a ditch but through a nozzle at high pressure. Aimed at the face of a cliff, the water could easily wash away tons of boulders, gravel, and dirt, and ounces of gold. The first use of hydraulic mining is credited to Edward Matteson in 1853, when he supplied the water through a rawhide hose to a nozzle he had carved out of wood. Later miners upgraded their hoses to canvas, and ultimately both the hoses (or pipes) and the nozzles became iron. Technological advances made the hose and nozzle connections more compatible and allowed for greater ease of movement and placement, and as the technology progressed, lavish attention was paid to the design and specifications of the nozzles, and multiple companies began producing competing appliances. The product names varied by company—Hoskin's Dictator and Hoskin's Little Giant are examples—but the name that stuck and became common parlance was the name the Craig Company had given its product: the Monitor.

Monitors were enormously powerful. In his multivolume classic *History of California*, historian Hubert Howe Bancroft wrote that an eight-inch monitor could throw 185,000 cubic feet of water in an hour with a velocity of 150 feet per second. Other accounts of the force are less technical but just as startling. One witness marveled that a strong man could not swing a crowbar through a six-inch monitor stream.

Yet another commented on the striking phenomenon of a fifty-pound boulder riding the crest of a jet with the power of a cannonball. Men were killed by the force of the water from two hundred feet away.

Water usage for every type of mining was mindboggling. Flumes, ditches, and aqueducts to service all mining operations in the region covered 5,726 miles. Rivers were diverted and streambeds dredged. In 1853, twenty-five miles of the Yuba River were redirected, and even longer stretches of the Feather and American Rivers were rerouted in 1855. But hydraulic mining required even more. At the North Bloomfield mine, sixty million gallons of water were used daily. Thomas Bell, the president of the company, estimated in 1876 that the hydraulic mine would consume sixteen *billion* gallons of water in that year alone.

The debris created was equally immense. North Bloomfield, a mine about one and a quarter miles long and 350 to 550 feet deep, excavated 41 million cubic yards of material between 1866 and 1884. Other hydraulic mines sprouted throughout the goldfields, and by 1891 federal government engineers estimated that all hydraulic mining had deposited 210,746,100 cubic yards of debris in the watersheds of just three rivers alone—the Yuba, the American, and the Bear. In 1928, a reassessment placed the figure even higher at 885 million cubic yards—more than three times the amount of earth moved in the construction of the Panama Canal.

The environmental effects were catastrophic. In 1859, Edward Vischer described the peculiar remnants of the process: "[There were] great masses of white limestone in bizarre shapes[;] . . . the earth, torn up everywhere, resembles a battlefield of the antidiluvian giants and monsters. Where the stone has been bared by taking away the earth covering, oblique ridges are revealed like irregular pillars or spires." Samuel Bowles, a visitor to the California goldfields in 1865, focused on the damage:

> Tornado, flood, earthquake and volcano combined could hardly make greater havoc, spread wider ruin and wreck, than are to be seen everywhere in the track of the larger gold-washing operations. . . . Thousands of acres of fine land along their banks are ruined forever by the deposits

> of this character. A farmer may have his whole estate turned into a barren waste by a flood of sand and gravel from some hydraulic mining up stream. . . . The tornout, dug-out . . . masses that have been or are being subjected to the hydraulics of the miners, are the very devil's chaos indeed. The country is full of them among the mining districts of the Sierra Nevada, and they are truly a terrible blot upon the face of Nature.

This "devil's chaos" was very profitable, but hydraulic mining was devastating to those downstream. In particular, two rivers—the Yuba and the Feather—were the main channels for the debris flow. Vast heaps of silt and mud, a.k.a. the slickens, crept slowly down the mountains on the riverbeds. When the debris hit the lowlands, the rivers deposited their silt in broad fans. Sand deposited at canyon mouths eventually swept out over nearby farmland, destroying it.

During the Gold Rush, the most environmentally destructive form of mining was hydraulic mining. Old Hilltop Mine at Michigan Bar, c. 1860. Courtesy of the California State Library, Sacramento; California History Section.

The neighboring towns of Marysville and Yuba City, at the confluence of the Yuba and Feather Rivers, were especially vulnerable. A cluster of large hydraulic operations upstream funneled vast amounts of slickens toward these valley communities, and by 1868 the beds of the two rivers stood higher than the streets, and a ring of levees as tall as the towns' rooftops were under construction. Boats found it impossible to navigate around the growing sandbars. The situation was intolerable and worsening daily.

But towns in the debris trail such as Marysville and Yuba City owed their prosperity and often their entire existence to the profitable mining companies upstream. The mines had existed before the towns had, and the federal government had patented their land, which the miners took to be an implicit approval of their operations. There were genuine questions of obligation, rights, and gratitude involved. Uncertain on how to proceed, Marysville and Yuba City tolerated the situation and did nothing for nearly two decades.

In 1872, farmers voiced their first organized opposition to the hydraulic mines and their destructive practices. The protests were mostly shouting into a hurricane, however. It quickly became obvious to the farmers that better organization was needed. And that would take some time.

Things came to a head in 1875. A flood in January broke the levees and dumped thousands of tons of mining debris into Marysville. It took most of the year to dig out from the mess that buried streets, homes, and businesses under yards of gravel and mud. Many buildings were damaged beyond repair and abandoned. Additional flooding later in the year undid much of the cleanup efforts, and tempers then reached a boiling point. A December meeting of farmers, lawyers, and business leaders meticulously chronicled their complaints and argued for a coordinated attack against the mining companies by lawsuits, legislation, or injunction.

On August 24, 1878, a determined throng of residents gathered at the courthouse in Yuba City. There, under the direction of James Keyes, the Anti-Debris Association of the Sacramento Valley was formed. The purpose of the association was to sue the individual hydraulic mining companies and the Hydraulic Miners Association over the use of public rivers and watersheds for deposit of slickens. Soon other anti-debris associations joined their ranks, and lawsuits piled up, demanding both compensation for damages and an end to the dumping of debris. However, the courts ruled that, because it was impossible to determine which particular mine was responsible for any degree of destruction, no specific damages could be determined or levied.

Endless litigation followed. In 1882, a lawsuit was filed in the United States Ninth Circuit Court under the name of Edwards Woodruff, the proprietor of a Marysville business block that had been flooded three times and the owner of two tracts of slickens-encrusted land in Yuba County. In reality, the lawsuit was brought by George Cadwalader, an attorney aligned with the anti-debris associations, who used Woodruff's name without his knowledge. This seminal case was entitled *Woodruff v. North Bloomfield Gravel Mining Company,* and the suit sought a "perpetual injunction" against the North Bloomfield Gravel and Mining Company and all other hydraulic operations on the Yuba River watershed for damaging farmland, commercial properties, and severely impeding river navigation.

The presiding judge was Lorenzo Sawyer, whom the mining companies felt would be favorable to their side. Sawyer had mined near Nevada City when he first arrived in Gold Country in 1849, and the mine owners felt assured he would understand their interests.

The jurist considered all viewpoints fairly and traveled throughout the region to examine the environmental impact of hydraulic mining. He visited numerous debris dams constructed to hold back the slickens and found them ineffective, collapsed, or buried under tons of debris. Some of the testimony he heard was startling. California's state engineer, William Hammond Hall, reported that more than 15,000 acres, or twenty-five square miles, of the Yuba River watershed was buried under mining debris, with the slickens at least twenty feet deep in the Yuba riverbed near Marysville. The court records noted one Dr. Eli Teegarden, who "owned 1,275 acres on the Yuba bottoms, some three or four miles above Marysville, on the north side. All except the 75 acres now lying outside the levee have been buried from three to five feet deep with sand, and utterly destroyed for farming purposes; for which injuries he has received no remuneration." There was the description of the collapse of the Yuba River's English Dam, which stored water for hydraulicking operations until, in 1883, it hurtled a torrent of water, at times ninety feet deep, as far as eighty-five miles from the Sierra Nevada to Marysville in about ten hours. There was also the surprising revelation that Lester Robinson, one of the owners of the North Bloomfield mine, had once sued a coal mine that had dumped tailings into a stream that fed onto, and then buried in debris, his San Joaquin Valley farm. Robinson won the suit and received monetary damages that were upheld under appeal.

After nearly eighteen months of weighing evidence, Judge Lorenzo Sawyer issued his decision in 1884. The mining companies, it turned out, had misjudged him. For the farmers: victory. For the hydraulic mining companies: unequivocal defeat. When printed, the decision, ever after known as "The Sawyer Decision," filled more than sixty pages of very small type in the legal record. But it was two little words that provided the greatest impact: "perpetual injunction."

Joy filled the streets of Sacramento, Wheatland, Chico, Red Bluff, Colusa, Stockton, and other valley farming towns affected by the slickens. Valley communities engaged in flag waving, shouting, shooting, drinking, and cannon blasts. Congratulatory messages were even delivered from the southern sections of the valley. On January 8, the *Marysville Daily Appeal* reported that Marysville's streets were "filled with people, shaking each other by the hand, cheering, and voicing their satisfaction and delight in the most hearty terms. The hand shaking and exchange of congratulations continued for hours. . . . At the same hour the church bells were ringing, and the steam whistles shrieked."

The hydraulic mining regions were not reveling—they were in mourning. The *San Francisco Daily Evening Bulletin* of January 8, 1884, reported a sense of "profound sorrow" in Grass Valley and "universal dissatisfaction and regret" in neighboring Nevada City. A Dutch Flat resident living near the North Bloomfield mine sadly announced that "most of us will pack our gripsacks" and depart.

The game was over. The hydraulic mining industry had been dealt a fatal blow. Any legal rationale for using rivers as slickens receptacles had been utterly and completely rejected. The conflicts and heartfelt tribulations over rights, responsibilities, and gratitude had been decided in favor of the farmers. The king was dead.

But that didn't mean the problems were solved. Ahead lay decades of expensive environmental cleanup, even as the fouled rivers continued to be contaminated with residual slickens from the closed mining operations, but those challenges were far from the thoughts of celebrants in the Sacramento Valley in January 1884. These were happy days, days of wondrous triumph over the "devil's chaos."

Woodruff v. North Bloomfield Gravel Mining Company led to the definitive demise of hydraulic mining during the Gold Rush, and within weeks mining companies folded their tents and withdrew, boomtowns disappeared, and unemployment swept through broad sections of Gold Country, although hard rock underground mining continued apace. But while the Sawyer Decision had brought an end to one industry, it gave birth to a new legal precedent: the principle that government has an important responsibility to prohibit severe environmental damage, even if it means the ruination of an entire industry. The legal standard of the Sawyer Decision meant that a grossly polluting industry could be shut down for the public good, and this case continues to be an important and frequently cited advance in environmental protection law.

Hydraulicking made a brief comeback with the passage of the 1893 Caminetti Act, which allowed for use of the technique under strict state regulation. The law featured demanding restrictions on debris flow, however, and since the industry could not be profitable due to exorbitant abatement costs, hydraulic mining disappeared for good in California after that single halfhearted revival.

"AS HUGE, TO ME, AS AN ELEPHANT"

GRIZZLY BEARS

As the first forty-niners plundered the goldfields, they sensed they were not alone. As they dashed through the land, grasping at anything shiny and filling their knapsacks, they felt something commanding and vicious prowling the streams and valleys that offered a profound threat to their existence. What ensued was a struggle for dominance of the landscape. It was a bloody battle between a remarkable force of nature and the insatiable power of greed. It is the story of the California Gold Rush versus the California grizzly.

The state was once a grizzly bear paradise. Nothing tangible remains in the wild of this noble beast today, but the ghosts of grizzlies still roam our dreams. We can imagine a world before the Gold Rush, before humans, when the fierce creature was the undisputed lord of the forest—"the sequoia of the animals," as celebrated naturalist John Muir rhapsodized in his 1901 classic *Our National Parks*. Even during the height of the Gold Rush, there was a sense that something majestic was being lost as the grizzlies began to vanish from the land.

In 1857, an article appeared in *Harper's New Monthly Magazine* entitled "The Grizzly Bear of California." The article is a litany of endless bear hunts and dramatic adventures with ursine giants, but it begins with this evocation of a time long past:

> Before the advent of Man, it was the Bear who asserted sovereignty over the animal and vegetable kingdom. But the king of bears reigned in California, where nature has

> made all things vast, extended, and overwhelming. Plains, over which the eye wearies itself with distances, green and interminable; a river with navigable arms, fed by all the snows of the Sierras, where large and solid streams plunge in unbroken falls over precipices thousands of feet in depth, into valleys where stand trees taller than cathedral spires.

But this "epoch of grandeur and strength" would change with the inexorable social deluge of the Gold Rush. Stories of scores of grizzlies playfully dancing in the moonlight and happily munching on manzanita berries—the so-called bear grapes—were replaced with tales of colossal, monstrous creatures menacing lives, livelihoods, and indeed the very future of the golden domain.

In the beginning, encounters with grizzlies (usually described as "interviews") were often mixtures of awe and threat, as was the case with the following incident from 1849. William Perkins, a merchant lately arrived from Canada, had ridden for two days south of his new home in Sonora, Tuolumne County, to collect some goods and now was making camp. While enjoying a pipe of "some genuine Virginia leaf," he heard a peculiar rustling sound at the cusp of the hollow, about a hundred feet distant. He raised his eyes to investigate and was met with the cause of the commotion. Thus began, as he noted, an "interview with a member of the Bruin family." At the edge of the clearing he spotted a grizzly bear pawing at an anthill and "licking up mouthfuls of the savoury insects with great apparent gusto." Perkins's heart raced and fear became an immediate and unwanted companion. Stories of unfortunate miners being maimed, mangled, and torn to shreds in bloody encounters with infuriated grizzlies were constantly floating around the goldfields, and now he himself spied, just ahead, "a huge bear, at least he appeared as huge, to me, as an elephant, but perhaps fear put a pair of magnifying glasses on my nose." Perkins did nothing to startle the bear or announce his own presence as he gently, slowly turned his head to look for an escape route. He spotted a nearby tree that he might easily climb if the bear, upon finishing his dinner of ants, attacked the trembling merchant. Perkins waited "in sufficient trepidation" for thirty minutes, although it felt like at least five hours, he remembered later. Presently, the grizzly, apparently uninterested in a dessert consisting of a Canadian forty-niner, meandered in the opposite direction of the camp and disappeared into the undergrowth. Understandably relieved, Perkins recollected, "I did not sleep much that night."

Grizzlies were abundant in those days. Horace Bell, a teenaged forty-niner who later became a militia ranger, a journalist, and a lawyer, declared that grizzlies were "more plentiful than pigs." Trapper and hunter George Yount, an associate of legendary mountain men Jedediah Smith and Jim Bridger, recorded that grizzlies "were everywhere—upon the plains, in the valleys, and on the mountains, venturing even within the camping grounds. . . . It was not unusual to see fifty or sixty within twenty-four hours." Jessie Benton Frémont, the extraordinary wife of explorer and statesman John C. Frémont, noted that the bears were frequent visitors to Mt. Bullion, a mining camp in the southern part of Gold Country. She wrote that it was a "famous resort for grizzly bears" and said "their 'wallows' were all around" and "it would not have been [surprising] to come back from a ride and find the Great Bear, the Middle-sized Bear and the Teeny-weeny-little Bear sitting in our chairs." Grizzlies were so numerous in 1848, one wag asserted, that the bears comprised the greatest population of large animals in the region, outnumbering the combined total of Californios, newly arriving Americans, and Native Californians.

And just as common as the grizzlies themselves were the stories about them. The historical record is brimming with chilling and thrilling descriptions of skirmishes with the bears. As time passed, seemingly every grizzly encountered was larger, hungrier, and more merciless and fearsome than the one before. In 1853, Joseph Wilkinson Hines, a gold seeker and missionary from Ohio, summarized the terrifying image of the grizzly:

> The most formidable and dreaded animal that roams the forests of California, or, indeed, of any other land on the face of the earth, is the grizzly bear. All other wild animals will flee from the presence of man, unless driven into a corner, or starved into desperation. But this shaggy monster roams

> the forests far and wide, seeking the weak and helpless of all classes as legitimate prey to his insatiable appetite. . . . Any one who has ever encountered one of these bloodthirsty brutes, especially when alone and destitute of deadly weapons or a friendly tree to climb, will, in all probability, never live to tell the sorrowful tale.

In 1856, the popular *Hutchings' Illustrated California Magazine* described the California grizzly as "the most formidable and ferocious of wild beasts." Exaggerated stories abounded of superhuman encounters with huge bears, twice or thrice the size of other species of bears. One gigantic specimen was said to exceed one ton—more than three times the standard weight. As *Hutchings' Illustrated* reported, a supposedly "typical" confrontation occurred in 1850 when a party of "six experienced hunters" spied a large grizzly along the banks of the American River:

> Hearing the crackling of bushes they immediately divided off in different directions, so as to surround [the bear]. At length he was seen, though partly hidden by the heavy underbrush, and fired upon, and at the first shot was badly wounded. This infuriated him, and he rushed quickly and suddenly out, and before the rifle could be re-loaded or the hunter (Mr. Wright) could escape, . . . he was tripped down, when the bear at one blow took out a piece of his skull to the brain, broke his arm, and would have been torn to pieces, but for the hasty advance of [others].

The grizzly was "shot through the heart" and then "another shot through the head laid him prostrate at [their] feet." Mr. Wright, "after several months of great suffering, eventually recovered."

Grizzly bear interactions sometimes occurred in clusters. Sonora merchant William Perkins documented an eventful week in August 1850 that included several such episodes within two or three miles of Sonora. As the week began, a hunter was assaulted by two bears and nearly killed. After a severe mauling, he escaped by climbing a tree. Four days later, a father and his two sons were attacked by a pair of bears. The father was torn to shreds and one son was gravely injured. The next day, a deer hunter was surrounded and assailed by eight angry bears. Seriously wounded, he played dead, and when the disinterested grizzlies finally sauntered away, he crawled to safety at a nearby ranch house. "The Grizzly Bear has come down from the mountains," Perkins wrote, "and are pretty abundant in the vicinity of Sonora, and make our deer hunting a dangerous sport."

Often the grizzlies were killed in groups, usually by guns and rifles but occasionally with poison or elaborate traps laced with liquor-drenched bait. Lee Summers Whipple-Haslam recalled that as a little girl in 1850s Tuolumne County, the bear hunting exploits of George Connally were renowned. Within a matter of minutes, she wrote, Connally dispatched four bears who were busily devouring a dead cow.

The fact that grizzlies sought to avoid human contact and conflict and were invariably described as peaceable when undisturbed was increasingly irrelevant to the pursuers, as the bear had come to exemplify more than prey. The California grizzly had become the formidable representation of an untamed landscape destined for subjugation by a new master. Grizzlies were powerful, independent, relentless, unpredictable, cunning, and insolent, and restraining or eliminating them required similar traits. That grizzlies were on the minds of many California settlers is reflected in profuse references to the bear in California place names; more than five hundred locations include "bear" as part of the name. There are seven Bear Rivers, twenty-five Bear Mountains, thirty Bear Canyons, and more than a hundred Bear Creeks. There are also waterways, meadows, and gulches sporting the names "Bearskin," "Bearpaw," and "Beartrap." Two hundred or more spots are labeled with "Grizzly," including several Grizzly Peaks throughout the state and the old mining site named Grizzly Flats in El Dorado County. At least a dozen locations are identified with "Oso," the Spanish word for bear. The grizzly adorns both California's state flag and its state seal. The mascots of several University of California campuses are bears.

The forty-niners may have gone to war with the bears to assert their own dominance, but there were practical applications, too. A growing population needed to be fed, and grizzly bear meat was one of the most readily available sources for market hunters. Gold Rush accounts often allude to meals of grizzly steaks or roast grizzly. Catherine Haun, newly arrived by wagon train to Sacramento in 1849, recalled that year's Christmas celebration: "I do not remember ever having had happier holiday times. For Christmas dinner we had a grizzly bear steak for which we paid $2.50." Bayard Taylor, a *New York Tribune* correspondent visiting the goldfields, reported in 1850 that "'grizzly bear steak,' became a choice dish at the eating houses. I had the satisfaction one night of eating a slice of one that had weighed eleven hundred pounds. The flesh was a bright red color, very solid, sweet, and nutritious; its flavor was preferable to that of the best pork."

Bear fat was also a desirable commodity. Called "bear grease," it was used for a variety of seemingly unrelated purposes. In *Our National Parks*, John Muir recalled a conversation with an old bear hunter:

> "B'ar meat," said a hunter from whom I was seeking information, "b'ar meat is the best meat in the mountains; their skins make the best beds, and their grease the best butter. Biscuit shortened with b'ar grease goes as far as beans; a man will walk all day on a couple of them biscuit."

Bear fat also became a treatment for thinning hair and baldness, as many believed that since grizzly bears were furry, their fat must have magical powers that could grow hair. (It did not.) Mixed with beef marrow, perfumed, and dyed green, bear grease was a popular hair pomade.

Bear remains were frequently spotted in market stalls throughout Gold Country. Balduin Möllhausen, a German artist, explorer, and topographer, described one such marketplace in 1850s San Francisco:

> [If a visitor] betakes himself to the game market, he might fancy himself in a zoological museum, so multifarious are the products of the chase heaped up before him. . . . In one [market] he sees . . . the great bear of the mountains, hanging to a proportionably strong hook, with dull dead eyes, but wide open bloody jaws, and countless heaps of hares, rabbits, and squirrels lying around him.

HUNTERS FIND A GRIZZLY.

Grizzly bears are now extinct in California, but they were common during the Gold Rush. "Hunters Find a Grizzly," *Pacific Rural Press,* March 1, 1873. Courtesy of the California State Library, Sacramento; California History Section.

Within a few years, however, grizzly sightings began to diminish. If the creatures were observed, it was generally in captivity, such as in carnival attraction or zoo exhibits. Bear skins surfaced as museum specimens or as decorative items such as rugs and furniture. The artist Thomas Hill displayed in his Yosemite studio the skin of the last grizzly killed in Yosemite Valley. United States presidents James Buchanan and Andrew Johnson were presented chairs fashioned from California grizzly bear carcasses. But there were fewer and fewer grizzly "interviews" in the fields and woodlands. The last wild California grizzly bear is believed to have been killed in 1908. These majestic monarchs of the forest had lived for thousands of years in the state, but it took humans just sixty years to stamp them out for good.

"A VERY NORMAL CHILDHOOD"

CHILDREN AND FAMILIES

Among the rarest of sights in the early days of the Gold Rush were families. Those with young children were particularly uncommon, and bound to draw sentimental deference from the forty-niners. The presence of women and children stirred suppressed emotions among the gold rushers, both those separated from their own wives and children as well as those who were only recently out of childhood themselves.

In March 1850, a mail steamer from Panama arrived in San Francisco carrying hundreds of passengers. As the vessel emptied, joyous reunions saturated the air with a mixture of laughter, weeping, hearty back slaps, and an electric atmosphere of excitement and anticipation. There were brothers reuniting with brothers, friends straining to see old comrades from back home, sons seeking fathers, fathers on the lookout for sons, and, as witness William White remembered, one young woman anxiously awaiting her appointment with the man who was to become her husband. White, a twenty-one-year-old businessman who had been in San Francisco only a few weeks, wrote, "She has come over the wild seas to her lover on an understanding that he is to meet her, with priest and witness, on the steamer deck and take her from there his wife." The bridegroom arrived, a hushed ceremony was performed before hundreds of onlookers, the groom kissed his bride, and the deed was done. But wait, there is more. The woman had in tow her four children, the oldest of which was an eight-year-old girl.

The Benson family—Lucy Emeline Strong Benson, baby Charles, and Henry Austin Benson—arrived in California in 1850 and lived in Hangtown, today's Placerville. Daguerreotype courtesy of the California State Library, Sacramento; California History Section.

The bridal party made its way down the gangplank and the multitude divided as if Moses were commanding the parting of the Red Sea. William White recollected the exhilaration:

> "Make way for the children! make way for the children! Stand aside! stand aside!" is now the cry from the crowd. Then some one calls out: "Three cheers for the children!" and "Three more for the mother who brought them!" adds another. . . . Oh! they are given with a will; for nothing in those days stirred the hearts of Californians as did the advent of a . . . woman and children.

Children likely reminded many men of families left behind, but they were also a symbol—a living, breathing manifestation—of a future in

California that most gold seekers did not intend to pursue. As New York argonaut Prentice Mulford wrote: "Five years at most was to be given to rifling California of her treasures, and then that country was to be thrown aside like a used-up newspaper and the rich adventurers would spend the remainder of their days in wealth, peace, and prosperity at their Eastern homes. No one talked then of going out 'to build up the glorious State of California.' No one then ever took any pride in the thought that he might be called a 'Californian.'" In the face of this, families were a sign that this perspective was shifting.

Theodore Barry and Benjamin Patten were both twenty-four years old when they settled in San Francisco in 1849. They witnessed many arrivals of hopeful argonauts, but their lasting recollections were of the greetings afforded children. They recalled the experience of one acquaintance disembarking in California:

> When he arrived in 1849, and walked up from the ship, with his wife and several little children, men crowded about the children, asking permission to kiss them, to shake hands with them, to give them gold specimens out of their chamois skin sacks, or little gold dust to make them rings, or something for an ornament. . . . The sight of their faces touched tender places in the hearts of men, divided by a continent's breadth from their own little ones; and to give other children toys, money, or something for their happiness, was a natural impulse.

Lewis Gunn was a teacher, printer, and antislavery activist in Philadelphia when the Gold Rush began. Bitten by the gold bug, Gunn left his home in 1849 to join the exodus to California, eventually settling in Sonora. By the end of 1849, he had been elected the recorder of Tuolumne County, and he began saving money to bring his wife, Elizabeth, and their four children to the Mother Lode. Two years later, Elizabeth and the children arrived in the Tuolumne County hamlet where Lewis published the *Sonora Herald* newspaper and owned a drugstore. A fifth child, Anna, was born in the town in 1853.

Elizabeth was apprehensive as she approached Sonora in 1851. She had heard stories of uncouth, even dangerous behavior from the rugged denizens of the rude mining camp, and she feared for the welfare of her children. Her concerns were allayed, however, as her entourage approached Knights Ferry on the stagecoach route to Sonora. In a letter written to her family back East on August 24, 1851, Elizabeth wrote that

> the gentlemen in the stage were much taken with the children; they bought them oranges and pears. At one stopping place the innkeeper . . . brought out a tumbler of milk and handed it to me for the children. The driver called "All aboard!" "No," said the [innkeeper], "not ready yet. Wait awhile, can't ye? Ye got to; the children want some milk." So it was handed to each of them. I was going to give it back, after the girls had had some, but they all insisted the boys should drink too. "Well," said the driver, "if it is for the children to drink milk, I'll stop."

While the goldfields were welcoming to children in many ways, it wasn't always the friendliest atmosphere in which to grow up. Anna Gunn Marston, the youngest daughter of Elizabeth and Lewis Gunn, recalled this gruesome scene from her childhood:

> [We saw] a wagon pass the house in which were four Chinamen sitting on long wooden boxes; and mother had to tell me that those were their coffins, and that they were being driven out of town to a place at some distance, where they were to be hanged for the murder of another Chinaman. Whenever there was to be a hanging, which happened not infrequently in those days, mother closed the house and tried to make us stay in rooms from which we could not see the crowds of men and women that streamed past to sit on the hillside overlooking the scene.

While children were frequently treated with kindness and care, their lot was never easy. Living in a new, unfamiliar landscape brimming with social and economic uncertainty, the potential of violence and the never-ending threat of fire and flood forced children to quickly adapt to difficult, unsettled circumstances. Sir Henry Huntley, a naval officer and a representative of a British gold mining company, wrote in his 1856 book *California: Its Gold and Its Inhabitants* that all the Gold Rush children he encountered were mature for their age; "I have never seen one act like a child."

But there was also some time for play and discovery in the eye of the social hurricane. Anna Gunn Marston recalled the "orderly and happy environment" her parents created in Sonora: "Our free outdoor play, our pets, the books father chose for us, the many simple celebrations mother planned, the love and spirit of co-operation which they put into our home life, gave us a very normal childhood." Her mother, Elizabeth Le Breton Gunn, remembered Christmas 1851 and the presents her four children received: cookies, candy, raisins, and, for some odd reason, a potato. She also recalled that each child received a piece of gold.

Marston's account of her upbringing in the Gold Rush opens a window to a world very different from the domain of the determined and frequently serious adults. Her memoir is filled with stories of wildflowers, pets, summer breezes, and stinky ointments. Anna remembered her mother's flower garden at the front of the house and how her father planted an apple tree for each of the five Gunn children. The kids often waded in an irrigating ditch that meandered through their property. "Our playground," Anna described, "extended to the high pine tree on top of the hill back of the house, beyond which we might not go. In springtime this hill was clothed with wild flowers of many kinds, among the loveliest of which were the deep rose-colored cyclamen (*Dodecatheon*), the mariposa lilies, and a fragrant low-growing white jasmine." Her brothers shot pine cones from the trees, and the girls would gather the pine nuts. On windy spring days, Anna's older brother Chester would "fly the huge kites that he made."

But perhaps her most enduring memory of childhood involved the discovery of a curious bottle. As she recalled, "On a shelf in the barn another child and I found a bottle of 'Mexican Mustang Liniment' and thought it would be nice to put some on our hair. The odor clung to us for a long time."

"THE YEARS HAVE BEEN FULL OF HARDSHIPS"

LUZENA STANLEY WILSON

In 1881, Luzena Stanley Wilson's youngest child, twenty-four-year-old Correnah, contracted a serious illness that required a long convalescence. To pass the time as her daughter recuperated, Luzena dictated to her the story of the family's journey to California and their time in the goldfields. The result was a longhand manuscript of about seventy pages. It is likely that Luzena reckoned that her narrative would never see the light of day, and, freed from any belief that her chronicle would become anything more than a handwritten family reminiscence stored in a trunk, the tale brims with an intriguing brew of anger, disappointment, awe, and pride very honestly rendered. Her accounts of loss, of death, and of constant rejuvenation are genuine and poignant.

The account began in 1848 Missouri:

> The gold excitement spread like wildfire, even out to our log cabin in the prairie, and as we had almost nothing to lose, and we might gain a fortune, we early caught the fever. My husband grew enthusiastic and wanted to start immediately, but I would not be left behind. I thought where he could go I could, and where I went I could take my two little toddling babies.

In 1849, when Luzena was thirty and her husband, Mason, was forty-four, the couple set out for the distant goldfields. As Luzena later told Correnah: "I little realized then the task I had undertaken. If I had, I think I should still be in my log cabin in Missouri. But when we talked it all over, it sounded like such a small task to go out to California, and once there fortune, of course, would come to us."

Two days into their trek, the Wilsons tried to join a larger wagon train called the Independence Company, which featured "five mule-teams, good wagons, banners flying, and a brass band playing." They were rebuffed, Luzena recalled, as the wagon train "'didn't want to be troubled with women and children; they were going to California'. My anger at their insulting answer roused my courage," she recalled, responding, "I am only a woman, . . . but I am going to California, too, and without the help of the Independence Co.!" Combining forces with some other hopeful travelers, the wagon train they formed was small by the usual standards—it had only six oxen-driven wagons—but it at least ensured they were "never alone." The Wilsons' company headed west in the spring of 1849 and soon encountered many other hopeful travelers bound for California:

> Ahead, as far as the eye could reach, a thin cloud of dust marked the route of the trains, and behind us, like the trail of a great serpent, it extended to the edge of civilization. The travelers were almost all men, but a mutual aim and a chivalric spirit in every heart raised up around me a host of friends, and not a man in the camp but would have screened me with his life from insult or injury . . . [on the journey to] my checkered life in the early days of California!

Luzena's depiction of the endless days on the trail, the boredom, the discomfort, and the backbreaking repetitiveness is succinct and evocative:

> Nothing but actual experience will give one an idea of the plodding, unvarying monotony, the vexations, the exhaustive energy, the throbs of hope, the depths of despair, through which we lived. Day after day, week after week, we went through the same weary routine of breaking camp at daybreak, yoking the oxen, cooking our meagre rations over a fire of sage-brush and scrub-oak; packing up again, coffee-pot and camp-kettle; washing our scanty wardrobe in the

> little streams we crossed; striking camp again at sunset, or later if wood and water were scarce. Tired, dusty, tried in temper, worn out in patience, we had to go over the weary experience tomorrow.

In September, their wagon company reached the crest of the Sierra Nevada. As she recalled later, "A more cheerful look came to every face; every step lightened; every heart beat with new aspirations. Already we began to forget the trials and hardships of the past, and to look forward with renewed hope to the future." As she cooked her family's supper on the campfire that first night on the western slope of the Sierra, a ravenous gold seeker approached her wagon. He was, Luzena noted, "attracted by the unusual sight of a woman." As he fished a gold coin from his pocket, the miner shyly said, "I'll give you five dollars, ma'am, for them biscuit." Five dollars was no small offering—the equivalent of $160 today—and she hesitated. The miner produced another gold coin, repeating his offer, and saying he would "give ten dollars for bread made by a woman [as he] laid the shining gold piece in my hand." Luzena realized what an exceedingly rare sight she was, a woman—with fresh biscuits!—in Gold Country in the early days of the rush.

She carefully deposited the precious coin in a box and placed the container in a safe spot in the wagon. However, the rough trail shook the box open and the coin fell to the ground, "hidden in the dust, miles back, up on the mountains. So we came, young, strong, healthy, hopeful, but penniless, into the new world," Luzena recalled with sadness.

The Wilsons' wagon company reached boomtown Sacramento in late 1849. The river city was lively and growing by leaps and bounds. The family purchased a canvas-walled hotel that consisted of three rooms—the kitchen, the guest quarters, and what Luzena dubbed the "general living room." The latter was, Luzena recalled, "a long room, dimly lighted by dripping tallow candles stuck in whisky bottles, with bunks built from floor to ceiling on either side" like shelves. Nestled in one corner of this warehouse hotel was a makeshift tavern staffed by a barkeeper, "the most important man in camp," dressed in "half sailor, half vaquero fashion, with a blue shirt rolled far back at the collar to display the snowy linen beneath, and his waist encircled by a flaming scarlet sash." Wedged into another corner was a perpetual card game infused with raucous cursing, a dance floor for half a dozen, and a fiddler furnishing music "under the manipulation of [his] rather clumsy fingers." But a grim veil of sadness shadowed the rollicking ambience Luzena described:

> One young man was reading a letter by a sputtering candle, and the tears rolling down his yet unbearded face told of the homesickness in his heart. Some of the men lay sick in their bunks, some lay asleep, and out from another bunk, upon this curious mingling of merriment and sadness stared the white face of a corpse. They had forgotten even to cover the still features with the edge of a blanket, and he lay there, in his rigid calmness, a silent unheeded witness to the acquired insensibility of the early settlers. What was one dead man, more or less! Nobody missed him. They would bury him tomorrow to make room for a new applicant for his bunk. The music and the dancing, the card-playing, drinking, and swearing went on unchecked by the hideous presence of Death. His face grew too familiar in those days to be a terror.

Just as they were in the goldfields, women were scarce on the streets of Sacramento. Luzena spent six months in the city and saw only two other women. The rambunctious young men of Sacramento were "a motley crowd," but whenever she entered the room

> the loud voices were hushed, the swearing ceased, the quarrels stopped, and deference and respect were as readily and as heartily tendered me as if I had been a queen. I was a queen. Any woman who had a womanly heart, who spoke a kindly, sympathetic word to the lonely, homesick men, was a queen, and lacked no honor which a subject could bestow.

The Wilsons began to earn their living, and then torrential rains arrived. For days and days, nonstop downpours assaulted Sacramento and inundated the low-lying community in the Great Flood of 1850. Luzena and Mason lost everything. For six weeks, the Wilsons' little hotel was awash in the floodwaters, and driftwood and the bloated carcasses of animals collected around the building's foundation. "In one corner lay our rusty stove," Luzena recalled, "the whole covered with slime and sediment. My husband . . . built a floating floor, which rose and sank with the tide, and at every footstep the water splashed up through the open cracks. We walked on a plank from the floor to the beds, under which hung great sheets of mould."

The couple vowed to escape Sacramento. Rumors began filtering into the devastated river city of a gold strike in the mining camp of Nevada City, fifty miles east in the Sierra foothills. The Wilsons desperately wished to join the flight to this new opportunity, and they hitched a ride with a sympathetic teamster, who accepted their promise to pay him the fare later. The roads were very muddy and slippery, and the wagon skidded constantly and crazily on the slick grades. Luzena

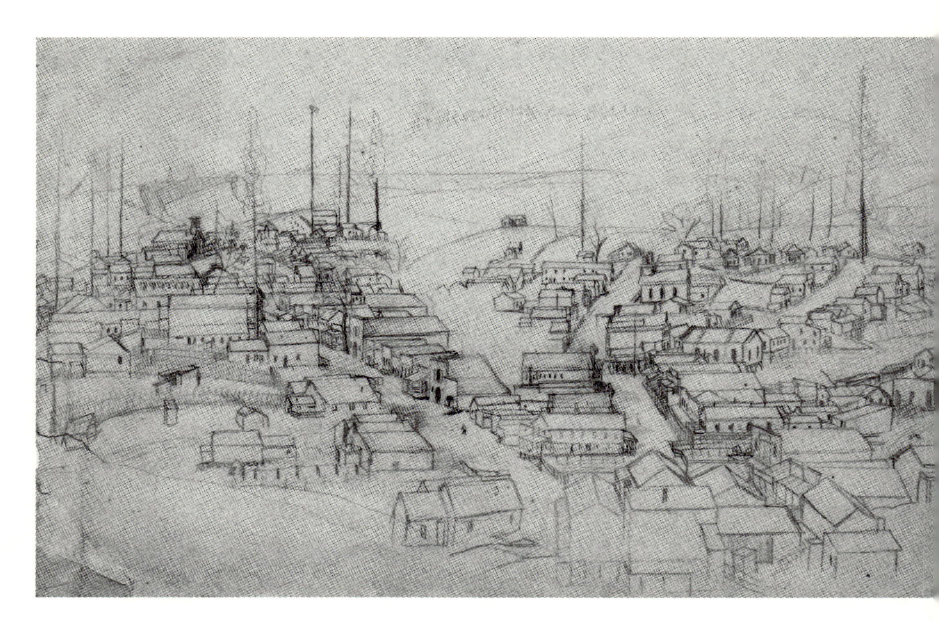

When Luzena Stanley Wilson arrived in Nevada City in 1850, the mining camp had become a boomtown. Nevada City, c. 1852, drawing by George Holbrook Baker. Courtesy of the California State Library, Sacramento; California History Section.

remembered, "Our wagon wheels looked like solid blocks; the color of the oxen was indistinguishable, and we were mud from head to foot." She recalled washing and scrubbing her young boys repeatedly until they returned "to their natural hue."

The Wilsons arrived to find Nevada City "a row of canvas tents lining each of the two ravines . . . flooded with the glory of the spring sunshine. . . . The gulches seemed alive with moving men. Great, brawny miners wielded the pick and shovel, while others stood knee deep in the icy water, and washed the soil from the gold."

Greatly impoverished by Sacramento's flood, Mason assembled a rough shelter of pine boughs and branches while Luzena began to consider "some plan to assist in the recuperation of the family finances." Already experienced in taking in boarders, and fully aware that the male population in mining camps had no desire to cook, clean, or do laundry, she decided to establish a new hotel right then and there. Nevada City already had one hotel, but she determined the booming town could support a second. She quickly crafted a dining table from two boards obtained from a generous neighbor and bought provisions at a neighboring store.

> When my husband came back at night he found . . . twenty miners eating at my table. Each man as he rose put a dollar in my hand and said I might count him as a permanent customer. I called my hotel 'El Dorado.'

The El Dorado Hotel was a success from its first moments, and within a few days Luzena realized they had enough money to compensate the kindly teamster who had offered them transportation to Nevada City. As the mining camp mushroomed, the Wilsons purchased adjacent property and added additional accommodations to their establishment. Within six weeks, Luzena proudly exclaimed, she had "from seventy-five to two hundred boarders at twenty-five dollars a week. I became luxurious and hired a cook and waiters."

Nevada City was expanding rapidly, and the hillsides sprouted haphazard clusters of canvas tents. The camp grew so quickly and changed appearance so dramatically that at one point Luzena had to ask directions to her own house. Soon there were more than ten thousand mostly male denizens of the mining camp. Daily life could be chaotic and dangerous. As Luzena noted:

> Men plunged wildly into every mode of dissipation to drown the homesickness so often gnawing at their hearts. They sang, danced, drank and caroused all night, and worked all day. They were possessed of the demon of recklessness, which always haunted the early mining camps. Blood was often shed, for a continual war raged between the miners and the gamblers.

But being one of the few women in town had its advantages:

> The feminine portion of the population was so small that there was no rivalry in dress or fashion, and every man thought every woman in that day a beauty. Even I have had men come forty miles over the mountains, just to look at me, and I never was called a handsome woman, in my best days, even by my most ardent admirers.

The Wilsons made an enormous amount of money very quickly. Six months after arriving in Nevada City, Luzena and Mason owned property and merchandise in excess of $40,000 (equal to $1.25 million today). Luzena kept hidden in her kitchen a jar in which she had squirreled away $500, and everything else she earned she invested back into the business. So busy and so profitable was their hotel enterprise, in fact, that they had no time to mine, including on a claim that was left to them by a miner who had struck it rich less than half a block from their hotel. The Wilsons sold the claim for $100; the purchaser proceeded to find $10,000 in gold.

Then tragedy struck. In March 1851, a careless resident set fire to a pile of pine shavings, and flames began a terrible dash through the camp. Within minutes, the canvas and wood city was fully engulfed. As Luzena sorrowfully reported:

> There were no means for stopping such a conflagration. Bells clanged and gongs sounded, but all to no purpose save

> to wake the sleeping people, for neither engines nor firemen were at hand. So we stood with bated breath, and watched the fiery monster crush in his great red jaws the homes we had toiled to build. The tinder-like pine houses ignited with a spark, and the fire raged and roared over the fated town. . . . The fire howled and moaned like a giant in an agony of pain, and the buildings crashed and fell as if he were striking them down in his writhings.

Eight thousand were homeless. Most were now destitute. Luzena believed that all was lost, but, in a stroke of luck, Mason had fortuitously placed the $500 from her jar into his pocket as the flames began licking at the hotel. It was all they had left. First flood, then fire: the Wilsons had tasted the bitter dregs of ruin twice since their arrival in California just two years earlier. The latest blow broke Luzena. Her health failed, and for weeks she was "bound in the fetters of fever."

When her condition improved, the Wilsons left Nevada City and took possession of a ramshackle hotel in downtown Sacramento but abandoned it after just a few weeks. As Luzena wryly noted, the hotel was "tenanted only by rats." They improvised shelter, ate when they could, and searched for a spot to begin anew. They found it near today's Vacaville.

> We were fascinated by the beauty of the little valley which already bore the name of Vaca from the Spanish owner of the grant within the limits of which it lay. The green hills smiled down on us through their sheeny veil of grass. The great oak trees, tall and stately, bent down their friendly arms as if to embrace us; the nodding oats sang a song of peace and plenty to the music of the soft wind; the inquisitive wild flowers, peeping up with round, wide opened eyes from the edge of every foot-path bade us stay. We made up our minds, if possible, to buy land and settle.

Luzena, Mason, and their now four children lived in the Vaca Valley for the next twenty-seven years. They established Wilson's Hotel, a successful and long-lasting venture. It was there that Luzena's daughter,

Correnah, recorded her mother's chronicle. In the last paragraph of her story, Luzena recalled, "The years have been full of hardships, but they have brought me many friends, and my memory of them is rich with pictures of their kind faces and echoes of their pleasant words." On the final page, Correnah added this postscript: "I have written my mother's story as nearly as I could in her words."

Correnah made a full recovery from her illness and kept the handwritten account of her mother's experiences in her possession. Years later she had two copies of the manuscript typed and bound. One of the books was donated to Mills College, where Correnah was the first alumnae trustee, and in 1937 the Mills College Library published Luzena Stanley Wilson's memoir. Curiously, despite the fact that more than fifty years had passed, the slim book was not supplemented with additional information about what had transpired in subsequent years or the ultimate fate of Luzena Stanley Wilson. Perhaps that is appropriate, as the indelible vision of this feisty Gold Rush pioneer, persevering against all odds, battling the elements, and ultimately finding her sanctuary, should remain the last image we remember of her.

SHADOW AND LIGHT

MIFFLIN WISTAR GIBBS AND DEFIANCE OF DISCRIMINATION

In early 1850, Mifflin Wistar Gibbs was in Rochester, New York. He was dejected and depressed. The twenty-seven-year-old free black man from Philadelphia had just completed a whirlwind tour throughout the eastern United States as an assistant to abolitionist Frederick Douglass. With Douglass's speaking tour finished and the excitement fading, Gibbs felt that his prospects were severely limited. As he later wrote, "The outlook for my future, to me, was not promising." Mifflin confided his gloom to Julia Griffiths, a British antislavery leader and a colleague of Douglass's. Griffiths responded with seven words that changed Gibbs's life: "What! Discouraged? Go do some great thing."

Armed with this powerful directive from Griffiths and a ticket provided by a sympathetic friend, Gibbs booked passage on the steamship *Golden Gate* and headed toward the distant realm of the dreamers. He had all his possessions in a trunk and sixty cents in his pocket. As Gibbs later noted in his autobiography, *Shadow and Light*, "Fortune . . . may sometime smile on the inert, but she seldom fails to surrender to pluck, tenacity and perseverance."

Mifflin Wistar Gibbs arrived in San Francisco in September 1850. After paying fifty cents to have his trunk delivered to his lodgings, he immediately spent his remaining few pennies on a cigar to celebrate

Mifflin Wistar Gibbs was a California Gold Rush civil rights pioneer, a successful merchant, and the first elected African American judge in United States history. Frontispiece from *Shadow and Light: An Autobiography, with Reminiscences of the Last and Present Century,* by Mifflin Wistar Gibbs (Washington, D.C.: 1902). Courtesy of the California State Library, Sacramento; California History Section.

his arrival in the land of golden opportunity. Over the next decade, his California experience would involve the full range of success and disappointment, exhilaration and exasperation, shadow and light.

Gibbs never seriously considered making a foray to the goldfields, preferring instead to remain in San Francisco. In the beginning of his time in the city, he was desperate for work and offered his services in the building trades. He was hired as a carpenter for $10 a day—nearly ten times the going rate for carpenters elsewhere in the United States—but that opportunity was short-lived. As he recalled in *Shadow and*

Light, "I was not allowed to long pursue carpentering. White employees finding me at work on the same building would 'strike.'" He worked briefly as a bootblack and then entered into a partnership operating a haberdashery with another African American man, Peter Lester. Gibbs and Lester made a fortune selling boots and men's clothing in boom-town San Francisco, and Gibbs rose from bootblack to prominent merchant and, later, to newspaper publisher. Despite his success, however, he could not escape the sharp sting of prejudice. A vicious incident involving his partner would place Mifflin Wistar Gibbs on the cutting edge of the effort to secure rights and respect for the African American community during the Gold Rush, and although Gibbs did not fully succeed, he never lost hope that his people would ultimately triumph over discrimination.

There were few African Americans in California during the Gold Rush. The notoriously inaccurate census of 1850 showed that the entire state was home to about one hundred black people that year, although the number rose significantly—to around twenty-five hundred, including ninety women, in 1852—as word spread that no other place in the United States offered greater opportunities for free blacks than California. But still their numbers remained low compared to other ethnic groups. By 1860 the census showed fifty-four hundred African Americans in California.

Although the state constitution had an antislavery proviso, that did not mean Gold Rush California was a sanctuary for African Americans. Especially galling to the black community, and particularly to Mifflin Wistar Gibbs, was the denial of the right to testify against white people in court. Gibbs saw his business partner, Peter Lester, robbed and beaten by a white thief, and as Gibbs was the only eyewitness but not allowed to testify, the assailant was not prosecuted.

Seeking equality before the law, petitions calling for the immediate removal of the testimony ban were circulated in African American enclaves, most notably in San Francisco. Newspaper editorials were published in support of the movement, hundreds of signatures were gathered, and the petitions were submitted, several times, to the State Assembly for consideration. The documents were repeatedly rejected.

By 1855, advocates in the black community were considering another, hopefully more fruitful tactic. They called for a state convention to formulate an activist response to the testimony ban.

In November 1855, the first in a series of "California Colored Conventions" assembled in Sacramento. The three-day convention was comprised of leading black citizens primarily from Northern California, and especially from San Francisco and Sacramento. Mifflin Wistar Gibbs was a prominent member and an official of the Committee on Credentials. The press cast a mocking eye on the proceedings, and when a grass fire broke near out near Sacramento on the first day of the convention, dropping ash on the city, the *Sacramento Daily Union* connected the fire with the opening of the Colored Convention: "A shower of ashes, proceeding from the burning tules of Yolo, fell upon the city and vicinity yesterday. . . . It was in reality a colored snowstorm, gotten up for the benefit of the Convention."

The delegates considered many pressing concerns but focused almost exclusively on the testimony question. As delegate David Lewis of Sacramento passionately explained:

> The law is to us a dead letter, a broken staff to lean upon. The oath that should protect life, liberty, and property, all that should throw the shield of law around ourselves and families, is denied us. Now we have no protection, and stand as nothing. "The oath" would make people careful how they act before us. We should have a voice. As it is, we are scarcely recognized as human beings.

The convention resolved to circulate more petitions and pledged to obtain numerous endorsements from white men—that is, people who had the right to vote. The outreach was a success, and petitions teeming with hundreds of signatures from empathetic white citizens were delivered to the California State Senate. Unlike the earlier documents, these were received by the legislature, but after that the matter was quickly buried in committee. But inertia was overcome and additional optimism began bubbling when a San Francisco grand jury report recommended that the testimony of non-white people be allowed.

A second Colored Convention was called for December 1856 in Sacramento. It was the largest gathering of its kind that California had ever seen. Gibbs was a member of the Executive Committee but remained in San Francisco to circulate petitions. Over four days, convention delegates once more concentrated on the testimony question. Building on earlier efforts, they authorized the circulation of additional petitions, and within weeks, signed petitions flooded the state capitol. The documents were accepted, yet again, by the state legislature, and, yet again, the question disappeared in the legislative labyrinth.

Many were discouraged as 1857 dawned, and a third convention gathered to rally supporters. That year also saw the dispiriting Dred Scott decision, in which the United States Supreme Court declared that African Americans had no status as legal citizens of the nation and, in the words of Chief Justice Roger B. Taney, were "beings of an inferior order, and altogether unfit to associate with the white race, either in social or political relations, and so inferior that they had no rights which the white man was bound to respect." Things would get worse for black Californians before they got better.

In 1857, as later court transcripts detailed, a white man named Charles Stovall brought a slave named Archy Lee to California and sold him to Robert Blakely. Lee fled from Blakely, who then had Lee arrested, and Blakely asked that his slave be returned to him under provisions of the federal Fugitive Slave Act of 1850. But United States Commissioner for California George Johnson determined that Archy Lee was not a fugitive under the meaning of the law and refused to order Lee back into Blakely's custody, instead ordering that he be released. The commissioner's reasoning was that Stovall had become a permanent resident of California and therefore could not own or sell a slave in a free state. In 1858 the Sacramento County Court upheld Johnson's decision, which was widely regarded by the black community and its supporters as a victory, as well as a much-needed, much-desired clarification of the status of African Americans in California.

Alas, the moment was short-lived.

The county decision was appealed to the California Supreme Court, where two justices, both Southerners with unmistakable pro-slavery sympathies, took it under advisement. One of them, Justice Peter Burnett, had been the first governor of the State of California and had previously tendered numerous anti-black pronouncements. For this case, he remained true to form and declared that Archy Lee was not free and should be reunited with his master Stovall without delay. Burnett argued that Stovall warranted an exception to the permanent-California-resident clause due to his inexperience and poor health and should be allowed to retain his slave despite the law clearly indicating otherwise. For black Californians, this judgment doused a long-awaited sunbeam of justice. Burnett's reasoning was a flimsy subversion and misinterpretation of well-defined California law, and his finding was roundly criticized not only in antislavery circles but in the legal community as well. Even Justice Burnett's colleagues on the supreme court condemned the finding. The *San Francisco Daily Alta California* called Burnett's opinion a "crowning absurdity and the greatest mass of legal contradiction that has ever come under our notice." The ruling was a disgrace, the newspaper added, and had rendered the California Supreme Court "a laughing stock in the eyes of the world." Within a month, the California Supreme Court's decision on Archy Lee was overturned by a federal district court in San Francisco, and although there were additional legal maneuverings, this federal court ruling stuck and Archy Lee was declared free, once and for all.

On March 19, 1858, emboldened by the Dred Scott decision, Nevada County assemblyman J. B. Warfield introduced Assembly Bill 339, titled "An Act to Restrict and Prevent the Immigration to and Residence in This State of Negroes and Mulattoes." It seemed likely to pass. Gibbs and editorial writers for black newspapers expressed outrage. Even though relatively few numbers of black people were immigrating to California, the act would have imposed severe restrictions on those who tried, and there were no protections for black individuals simply visiting the state for business, pleasure, or family or personal matters. Under the proposed law, black citizens would have to register to prove their residency and would be required to carry their registration certificates at all times or risk fines and imprisonment. This type of regulation was not compulsory for any other group in the state. Surprisingly, due to technicalities the measure did not pass, but its near approval proved to be the final straw for hundreds of black Californians. They had endured enough and decided it was time to leave the Golden State.

Several hundred African Americans sought a new home where they would find tolerance, opportunity, and acceptance. This "promised land" appeared to be sparsely populated British Columbia, which was in the early days of a gold rush of its own. In April 1858, a vanguard of two hundred embarked on an exodus to Victoria, British Columbia, aboard the sailing ship *Commodore*. Among its members was Archy Lee. Mifflin Wistar Gibbs offered a farewell address and was seriously considering joining the migration himself, as were hundreds of others.

As positive word filtered back from the Victoria migrants, plans were accelerated for additional settlers to head north. Ultimately, more than five hundred would join the early arrivals in Victoria and Vancouver. Nearly 15 percent of all black people in California migrated to British Columbia during this period, and the *San Francisco Daily Evening Bulletin* predicted that "the day when colored people fled persecution in California . . . may yet be celebrated in story." A conference of ministers for the African Methodist Episcopal (A.M.E.) Church issued a resolution that referred to the Victoria mission as "God's rescue."

In June 1858, Mifflin Wistar Gibbs left San Francisco for British Columbia. He carried with him stores of supplies needed by the settlement and, upon arrival in Victoria, immediately flourished there as a merchant. The community was succeeding, too. Letters home described that nearly all the jobs in Victoria were held by black people; that the migrants had purchased considerable tracts of land; and that the police force—the only police force—was the black unit known as the Victoria Pioneer Rifle Corps. It was a step up from conditions in California, but it was not idyllic. Many white Americans had rushed to British Columbia for its gold, and they harbored the same familiar prejudices; tension between these old adversaries was inevitable. There were a few efforts to segregate churches, theaters, and saloons, but, in general, these new black Canadians found support and protection from both the British government and local clergy.

Mifflin Wistar Gibbs sunk roots in the new colony and became one of the wealthiest men in Victoria. His five children were born in British Columbia, and he was elected to the Victoria City Council. But even then he could not escape intolerance and social indignities. In 1861, while attending the theater, a white bigot objected to sitting in the same section with Gibbs and threw a container of flour at him and his wife.

When the Civil War ended, prospects for African Americans in the United States initially appeared brighter. While the members of the exodus to B.C. had pledged allegiance to the British crown, many retained unbroken loyalties to American family and friends, as well as to the United States, and in the postwar years there was a reverse migration. Hundreds of black Americans remained in Canada, but a significant portion of the migrant community ended their exile and returned to the United States, where they resumed their former lives. Among those starting anew in his former homeland was Mifflin Wistar Gibbs.

After more than ten years in Victoria, Gibbs migrated to Ohio, where he obtained a law degree at the age of fifty. After hearing of promising potential in Arkansas, he moved to Little Rock in 1871 and hung up his legal shingle. Within two years, Gibbs was appointed County Attorney in Pulaski County, and, in 1873, he was elected Municipal Judge in Little Rock. His election generated national interest, as the majority of the voters were white and he was the first elected African American judge in United States history. In 1877, Gibbs was appointed the Registrar of United States Lands in the Little Rock District by President Rutherford B. Hayes. Gibbs received another federal appointment from President Benjamin Harrison in 1889, and in 1897 Judge Gibbs was selected by President William McKinley as United States Consul for Madagascar.

Mifflin Wistar Gibbs died in 1915 at the age of ninety-two. His passing came sixty-five years after abolitionist Julia Griffiths had urged the nearly penniless young man to "Go do some great thing." It is likely that Griffiths would have felt that Mifflin Wistar Gibbs—California Gold Rush civil rights champion, defiant migrant, successful merchant, respected lawyer, city council member, county attorney, judge, federal official, and ambassador—had succeeded.

A VAST, GLOWING EMPTY PAGE

REINVENTION AND THE VERITABLE SQUIBOB

Virtually everyone who came to California donned a different guise for the adventure. Farmers became sailors, teamsters wielded picks and shovels, doctors sold dry goods, clerks transfigured into carpenters, students became stevedores, and almost everyone transformed into miners, if only briefly. Social reformer Eliza Farnham captured this California conversion when she wrote the following in 1856:

> If [a new arrival] could blow a fife on training days, he will be a professor of music here; if he have built a pig-sty or kennel at home, he will be a master-builder in California. . . . He has a wide range of pursuits, places, and employments to choose from. . . . He may keep a monte table, sell strong drink, be treasurer of moneyed associations, or quartz companies, in short, he may be anything that he has the power or the wish to be.

This opportunity to reinvent oneself also came in handy following the disappointments and failures that were a part of so many ventures in Gold Rush California. James S. Holliday, author of the seminal 1981 history *The World Rushed In*, wrote that the Gold Rush was an environment in which "failure was commonplace, almost a given, a part of their way of life, [and] of the process of seeking success." With "no hometown eyes" watching and no critical dismissal accompanying this failure, however, personal rebirth and reinvention was recurrent and familiar, and, for many, the opportunity to dust oneself off and rise again helped lessen the sting of failure, provide insulation against disillusionment and despair, or simply offer a license for intemperance that would be scorned back home.

And yet trying on a new persona was not always comfortable. Sometimes it even necessitated the creation of a false identity. Overall, it helped to have a sense of humor, as did a handful of writers who became famous for capturing the California spirit from a unique angle.

While most of the writing that came out of California during this time was nonfiction, there were a few chroniclers coping with the Gold Rush experience in an oddball fashion—particularly those who transformed very real concerns into satirical commentary. From the earliest days, writers deflected worry into whimsy, and song lyrics in particular were frequently humorous observations on the trials and tribulations of the moment. Whether on the trail, aboard a ship, or in the mining camps, a handful of writers and composers described the events with a humorous bent—from genial to sardonic—and a few of them regularly reinvented themselves in the process as literary or musical lights. Through these new personas, and almost always under pseudonyms, they created a genre of their own, "California Humor," a distinctive form distinguished by madcap overstatement, tongue-twisting names, comical author aliases, and exaggerated, absurdist observations puncturing the pretenses of daily life. The style still resonates in satire and comedy today.

Among the most famous practitioners were humorist Alonzo Delano, Grass Valley's first city treasurer, who wrote under the pen name Old Block, and, most famous of all, Samuel Clemens, better known as Mark Twain. Others who have now faded into obscurity but were popular during the era include James and Donald Read, who wrote *Journey to the Gold Diggins* (1849) under the name "Jeremiah Saddlebags"; George Washington Peck, with his fantasy *Aurifodina* (1849), using the pseudonym "Cantell A. Bigly"; the mysterious Brother Jonathan, who penned *The Fortunes of Ferdinand Flipper* (1850); and the author identified only as X O X, who tendered *An Outline History of an Expedition to California: Containing the Fate of the Get All You Can Mining Association* (1849).

Writers of comic songs were especially popular, even when they were anonymous. The unknown lyricist of "The Happy Miner" offered this touching thought:

> No matter whether rich or poor
> I'm happy as a clam.
> I wish my friends at home could look
> And see me as I am.
> With woolen shirt and rubber boots
> In mud up to my knees,
> And lice as big as chili beans
> A-fighting with the fleas.

But among the personalities that populated the field, it was an acknowledged pioneer of the genre whose secret identity was most surprising. Few remember him today, but this very popular Gold Rush writer was known as "John Phoenix" or, most famously, "The Veritable Squibob." He first came to public notice because of fleas.

The undeniable king of the animals during the Gold Rush was the mighty grizzly bear, but the most ubiquitous critter was the diminutive acquaintance of many a forty-niner: the lowly flea. The era's accounts are rife with lamentations about the irritating pest. Tireless evangelist William Taylor characterized the vermin as the "liliputian host of the flea tribe," and New York gold seeker Hiram Pierce stated the obvious: "The fleas hold undisputed Sway." So common were fleas that most gold rushers attempted to battle them using a mind-over-matter technique, which either meant pointless attempts at ignoring them completely or simply adjusting to the scratchy reality. William Taylor recalled his conversation with a fellow minister, the Reverend Mr. Trumbull, who recounted, "When I first came to this place I feared the fleas would worry the life out of me. I could neither eat nor sleep, nor stay awake with any comfort. But after a few weeks I got used to them, and now I pay no attention to them. The biting of a dozen at once don't cause me to wince, nor lift my pen from my paper." Most did not possess the willpower of Trumbull, however, and persistently sought some relief from the aggravating infestation.

Enter the perpetually helpful and off-kilter newspaper correspondent John Phoenix with his "antidote" for fleas. "In a climate where the attacks of fleas are a constant source of annoyance," he opined, "any method which will alleviate them becomes a *desideratum*." The solution is a simple recipe, which Phoenix wrote with tongue firmly in cheek:

> Boil a quart of tar until it becomes quite thin. Remove the clothing, and before the tar becomes perfectly cool, with a broad flat brush apply a thin, smooth coating to the entire surface of the body and limbs. While the tar remains soft, the flea becomes entangled in its tenacious folds, and is rendered perfectly harmless; but it will soon form a hard, smooth coating, entirely impervious to his bite. Should the coating crack at the knee or elbow-joints, it is merely necessary to retouch it slightly at those places.

The remedy is effective, simple, and inexpensive, John Phoenix continued, and should "be removed every three or four weeks."

If wearing a tar suit for a month did not work, Phoenix offered an even simpler alternative: "On feeling the bite of a flea, thrust the part bitten immediately into boiling water. The heat of the water destroys the insect and instantly removes the pain of the bite." Of course, he noted, he had not tried this himself; it was a promising treatment "in theory only."

This charming mid-nineteenth-century fluff is especially interesting when you discover the true identity of John Phoenix. One would imagine a clever fellow, a jovial quipster and *bon vivant*, composing his satirical passages for the amusement of his private club. But John Phoenix was actually a professional military officer named Lieutenant George Horatio Derby, a topographical engineer with the United States Army.

Derby had always been unpredictable—a scamp and a rebel. Born into a leading family in Massachusetts in 1823, he was expelled from a snooty Boston boarding school at the age of twelve for what was characterized as his "native relish for sin." In 1842 he entered West Point, where he quickly gained a reputation for outlandish pranks and a wicked sense of humor. Try as they might, the military academy could

Lieutenant George Horatio Derby: Gold Rush topographical engineer and a humorist otherwise known as the Veritable Squibob. University of California, Berkeley; Bancroft Library; Derby, George Horatio–POR2.

---

not rein in his antics, but despite his love of a good time, Derby took his studies seriously and, in 1846, he graduated seventh in his class at West Point, ranking above classmates and future Civil War generals George McClellan and Thomas "Stonewall" Jackson. Following graduation, Derby joined the United States Army Corps of Topographical Engineers, and the next year, during the Mexican War, he was seriously wounded at the Battle of Cerro Gordo. In 1849, he was assigned to explore and survey California and the Colorado River. During his tour, Derby happily participated in amateur theatricals and supplemented his income by publishing humorous vignettes and commentaries in San Francisco's *Daily Alta California*.

His imaginative and often silly articles were very popular, and his writings were frequently reprinted in newspapers throughout the United States. Among his many articles, Derby, as Squibob, described the unbearable heat in the Arizona desert and added, "One of our Fort Yuma men died, and unfortunately went to hell. He wasn't there one day before he telegraphed for his blankets." About the Oregon Territory, through which many gold seekers had traveled to reach California, he commented that "it rains incessantly twenty-six hours a day for seventeen months of the year." Squibob also offered "A New System of English Grammar," which replaced adjectives with numbers; regarding this "glorious, soul-inspiring idea," he wrote, "Let *perfection* . . . be represented by 100, and an absolute minimum of all qualities by the number 1. Then by applying the numbers between, to the adjectives used in conversation, we shall be able to arrive at a very close approximation to the idea we wish to convey." The system would allow the public to become "at once an exact, precise, mathematical, truth-telling people. It will apply to everything but politics; there, truth being of no account, the system is useless. But in literature, how admirable!" Squibob obligingly provided an example with his description of a chance encounter on the street:

> As a 19 young and 76 beautiful lady was 52 gayly tripping down the sidewalk of our 84 frequented street, she accidentally came in contact—100 (this shows that she came in close contact)—with a 73 fat, but 87 good-humored-looking gentleman, who was 93 (i.e., intently) gazing into the window of a toyshop. Gracefully 56 extricating herself, she received the excuses of the 96 embarrassed Falstaff with a 68 bland smile, and continued on her way.

On another occasion, in the interest of historical documentation, the Veritable Squibob provided his readership with a typical "California Love Letter" from a pining argonaut to his long-suffering sweetheart. It read, in part:

> Mariesville july fore, 1856

> Dear Cate, you know I luv you mor an any other Girle in the World, and wat's the Reson you allways want Me to tell you so. I no you ar almost gitting tired of waiting for me; I no you luv me fit to brake your hart. . . . I no I out to marid long ago. Leven years is rather long to kort a gal, but ile hav you yit Cate.
>
> Good by, till next we meet,
> Your affeckunate Lover,
> D—— G——

Even with these outlets, however, Derby could not restrain his offbeat wit and flippant attitude during his official duties, a circumstance that irritated his superiors and occasionally led to reprimand. For instance, in 1850, Lt. Derby of the Topographical Engineers of the United States Army provided a notable section of a congressional monograph entitled *A Report of the Secretary of War Communicating Information in Relation to the Geology and Topography of California*. Derby's "topographical memoir" of an 1849 journey in the Sacramento Valley is mostly serious, and yet it is sprinkled with the author's usual glib asides. Derby described how "the mules refused to pull, until . . . by dint of much shouting, screaming and profanity from the party, they managed to draw the empty wagon to the top." When two of his men refused to work on a Sunday, Lt. Derby refused to pay them, and he reports that, "upon hearing this, their devotional feelings subsided with vast rapidity." Not content with simply saying he had seen raccoons, they became "three veritable raccoons"; rather than dryly noting that he encountered an emigrant wagon train, Derby added that it contained "unhappy-looking women and unwholesome children." The lieutenant's comments did not please Secretary of War George W. Crawford, and Derby was chastised.

The Veritable Squibob largely disappeared after George Horatio Derby left California in 1856, assigned to supervise the construction of lighthouses in Alabama. During those later years, he suffered from vision loss, believed to have been caused by a brain tumor, and as Derby gradually lost the ability to read and write, he grew increasingly downhearted. He died in 1861 at the age of thirty-eight.

"EMPEROR OF THESE UNITED STATES"

JOSHUA NORTON

In October 1865, Samuel Clemens, about to become world famous as Mark Twain, wrote to his older brother, Orion, formerly Secretary of the Nevada Territory, in Carson City. Recounting the twisting, shifting paths his own life had followed, Sam evoked an old acquaintance in San Francisco: "It is human nature to yearn to be what we were never intended for. It is singular, but it is so. I wanted to be a pilot or a preacher, & I was about as well calculated for either as is poor Emperor Norton for Chief Justice of the United States." That Samuel Clemens chose Emperor Norton to exemplify the trials and quirks of personal transformation is not surprising. Perhaps the best-known example of Gold Rush reinvention is the tale of Joshua Abraham Norton, also known as Norton I, Emperor of the United States and Protector of Mexico. His is a story of rebirth constructed partly from reality and largely from whole cloth—a chronicle of failure, survival, intentional manipulation, and cynical exploitation, all exhibited in one vibrant package.

Joshua Abraham Norton was born in England in 1818 and was raised in South Africa. His parents, John and Sarah Norton, were successful maritime outfitters and leaders in their Jewish community. When the Gold Rush began, Joshua found himself in Cape Town, a failed business under his belt and with no family after the deaths of his parents and brothers. He had inherited his father's estate, estimated at $40,000 (equal to about $1.2 million today) and departed for California in 1849.

When he arrived in San Francisco he signed the hotel registry as "Joshua Abraham Norton, International Merchant." He set himself up in real estate and traded commodities, and he was very successful at both. He purchased real estate downtown, bought a ship anchored in the harbor for a warehouse, and secured offices in a stately granite structure that also housed the British Consulate. Within a few years of his arrival, Joshua Norton was a valued member of many influential social clubs and civic organizations, and by 1852, his assets exceeded $250,000, a value of about $8 million today.

And then financial disaster.

In 1853, Norton attempted to corner the market in rice during a famine in China that had cut off exports to the West. He entered into a contract, made a large down payment on a shipment of Peruvian rice, and promised to pay the balance within thirty days. Not long after, boatloads of rice began arriving in San Francisco, prices plummeted, and Norton, whose rice was found to be of inferior quality, attempted to nullify his contract. He was sued, accused of embezzlement, and tormented by colossal legal bills. Norton lost his case and was forced to sell his real estate and businesses at a loss to satisfy his creditors. By 1856 he was broke. His memberships in private clubs were revoked, and he filed for relief under Insolvency Law. In 1858, Norton was listed in the San Francisco city directory as residing in a low-rent, working-class boarding house in a neighborhood a few blocks from the Financial District where he once held sway.

Joshua Norton would have remained merely another anonymous casualty of the economic rollercoaster that was Gold Rush California except for a curious incident at the office of the *San Francisco Bulletin* on September 17, 1859. "The world is full of queer people," George Fitch, the *Bulletin*'s editor, recalled. "This forenoon, a well-dressed and serious-looking man entered our office, and quietly left the following document, which he respectfully requested we examine and insert in the *Bulletin*. Promising him we would look at it, he politely retired, not saying anything further." Published the next day in the *Bulletin*, the document was a proclamation announcing the ascendency of Norton I, Emperor of the United States. It read:

> At the pre-emptory request of a large majority of the citizens of these United States, I Joshua Norton, formerly of Algoa Bay, Cape of Good Hope, and now for the last nine years and ten months past of San Francisco, California, declare and proclaim myself the Emperor of These United States, and in virtue of the authority thereby in me vested do hereby order and direct the representatives of the different States of the Union to assemble in Musical Hall of this city, on the 1st day of February next, then and there to make such alterations in the existing laws of the Union as may ameliorate the evils under which the country is laboring, and thereby cause confidence to exist, both at home and abroad, in our stability and integrity.
>
> Norton I, Emperor of the United States

The reign of Norton I had begun. It would continue for the next twenty-one years in all its magnificent and melancholy tragicomedy. Readers clamored for more proclamations, and editor George Fitch saw an immediate rise in the *Bulletin*'s circulation. Fitch encouraged Norton to continue his pronouncements, and, in rapid fire, the emperor began issuing decrees and garnering wider notice. Some viewed him as a delightful eccentric, a harmless crank, but others echoed the sentiment of English writer and San Francisco resident Walter Fisher, who dismissed Norton as "a crazy beggar." Either way, the public was fascinated. They were particularly taken with his personal appearance and unique clothing. The Reverend O. P. Fitzgerald, who counted Norton as a regular and frequently judgmental visitor to his church services, offer this description:

> The Emperor Norton—THAT was his title. He wore it with an air that was a strange mixture of the mock-heroic and the pathetic. . . . Arrayed in a faded-blue uniform, with brass buttons and epaulettes, wearing a cocked-hat with an eagle's feather, and at times with a rusty sword at his side, he was a conspicuous figure in the streets of San Francisco, and a regular *habitué* of all its public places. In person he was stout, full-chested, though slightly stooped, with a large

> head heavily coated with bushy black hair, an aquiline nose, and dark gray eyes, whose mild expression added to the benignity of his face. On the end of his nose grew a tuft of long hairs, which he seemed to prize as a natural mark of royalty, or chieftainship.

In his declarations, all duly published in the many rival San Francisco newspapers, Norton abolished Congress, dissolved the California Supreme Court, and fired Virginia governor Henry Wise. When Congress later met despite his abolition command, Norton I ordered the "Commander-in-Chief of the Armies" to "clear the halls on Congress." Following Napoleon III's invasion of Mexico in 1863, Emperor Norton added to his title, declaring he would forever afterward be referred to as "Norton I, Emperor of the United States and Protector of Mexico."

Norton energetically sustained his efforts to transform the United States to meet his whims and desires, and on the eve of the Civil War, he ordered the dissolution of the United States government. His proclamation read: "We are certain that nothing will save the nation from utter ruin except an absolute monarchy under the supervision and authority of an independent Emperor"—meaning Emperor Norton, of course. Along a similar path, Norton would also abolish the Democratic and Republican political parties. In 1872, he ordered a suspension bridge be constructed across San Francisco Bay to Oakland, a vision that predated the construction of the Bay Bridge by more than sixty years. Also showing particular foresight, Norton railed against those referring to San Francisco as "Frisco," decreeing that anyone so doing "shall be deemed guilty of a High Misdemeanor, and shall pay into the Imperial Treasury as penalty the sum of twenty-five dollars."

San Franciscans embraced Emperor Norton, to a degree. He rode free on all city transportation, a perk that was later extended to California railroads via a pass issued by railroad baron Leland Stanford. City police saluted the emperor as he passed, and politicians wooed Norton and openly used him to push their political agendas. Restaurants provided free meals to the emperor in return for advertised endorsements from him, and newspaper reporters were known to concoct Emperor Norton stories and to print bogus proclamations as a way of selling more newspapers in the city's heated publishing wars. Merchants discovered they could receive free publicity if they conceived and promoted some tenuous connection to Norton, and over the years, various businesses reaped big profits from Emperor Norton dolls, complete with plumed hats; postcards and lithographs printed with Norton's image; and even a fashionable Emperor Norton cigar. Published accounts stated that Norton was always accompanied by two dogs, his beloved Bummer and Lazarus. This was a myth, however, as the creatures were not Norton's pets but two stray rat-killing dogs that had been adopted by the board of supervisors as official San Francisco mascots. Reports indicated that Norton thoroughly disliked being associated with the animals, and he didn't take kindly to a series of popular lithographs featuring himself and the dogs. Produced by artist Edward Jump in the early 1860s, the drawings were titled "The Three Bummers"—"bummer" being a nineteenth-century term for an idle loafer. The emperor was not amused, and upon seeing one of the prints in a store window, he smashed the glass with his cane. Rarely did any benefit accrue to Joshua Norton beyond his first contact, beyond the first exploitation of his persona.

Despite his notoriety, Norton survived primarily due to the kindness of old associates and strangers. Unlike the commonly accepted fairytale that the emperor's existence was an endless carnival of free dinners, bottomless drinks, and blissful encounters with an adoring public, his life was grim. He wore tattered, threadbare hand-me-down uniforms provided by the army's Presidio, and as he got older he scrupulously alternated wearing Union and Confederate uniforms so as not to offend either side. As his fortunes continued to decline, Norton moved to a fifty-four-square-foot room in a grimy building called the Eureka Lodgings on Commercial Street. This room was his home for nearly twenty years. He hung his uniforms on ten-penny nails. Emperor Norton reluctantly accepted the occasional handful of coins to pay for meals or cover his rent, but to save embarrassment he described the handouts as "taxes" or distributed valueless "Imperial" scrip in return. It was during this time that *San Francisco Call* reporter Samuel Clemens encountered the emperor, recalling in 1890, "O dear,

Joshua Abraham Norton, self-styled as Norton I, Emperor of the United States and Protector of Mexico. Courtesy of the California State Library, Sacramento; California History Section.

it was always a painful thing for me to see the Emperor begging, for although nobody else believed he was an emperor, he believed it."

On January 21, 1867, an overenthusiastic rookie policemen arrested Norton on charges of vagrancy and "lunacy." San Francisco sprang to the emperor's defense. Typical was this comment from the *Alta California* on January 22: "He has shed no blood, robbed nobody, and despoiled the country of no one, which is more than can be said of any of his fellows in that line." San Francisco's chief of police, Patrick Crowley, released Norton and issued a public apology. In 1870, the United States Census listed Norton I as "insane," but those who knew him said that he was always lucid, conversant on many topics—most notably history and science—and an excellent chess player. He was in many ways a walking, talking paradox.

On January 8, 1880, Emperor Norton died after collapsing on a San Francisco street. His body was taken to the city morgue, and in his pockets was found $5.50, an 1828 French franc, and ninety-eight shares of worthless gold mine stock. In his raggedy coat were a handful of crumpled telegrams from Czar Alexander II of Russia, England's Queen Victoria, and the president of the French Republic. They were all addressed to the emperor, but they were also all fakes, poking fun at Norton. Obituaries for Emperor Norton were published throughout the United States; the *San Francisco Alta California* devoted thirty-four column inches to Norton's passing, the *Cincinnati Enquirer* offered sixteen column inches, and the *Sacramento Daily Union* offered this remembrance on January 10, 1880:

> [Norton] was a perfectly harmless, innocent, and good-natured old man, and his mild absurdities amused a community which was not disposed to examine his credentials too critically. Even if he was perfectly sane he might have been in much worse business, for he might have been a tonguey demagogue, misleading the people, living in idleness on the earnings of others, and spreading a moral pestilence among his generation. Emperor Norton was at least a more respectable character than one of these creatures, and his memory is infinitely more deserving of preservation.

More than ten thousand mourners passed Norton's coffin as it lay in state at the San Francisco city morgue. James Eastland, a wealthy businessman who knew Joshua Abraham Norton as a merchant and member of the Freemasons during the Gold Rush, collected donations for a proper funeral for his destitute old associate. The funeral cortege to the cemetery strung out for more than two miles. As the interment ceremony unfolded, the sky darkened from a total eclipse of the sun.

In print, the Reverend O. P. Fitzgerald concluded that Norton's life was "an illusion." But so were the lives of many reinvented personalities during the Gold Rush. For most, the illusion was brief. The story of Norton I, Emperor of the United States and Protector of Mexico, has lasted longer.

"VERY LITTLE LAW OF ANY KIND"

LAWYERS AND JUDGES

Starting with one wagon and a mule, teamster John McGlynn prospered during the giddy early days of the Gold Rush in San Francisco. As time passed, his business grew and he purchased a second wagon, but now he needed someone to drive it. He wrote to his mother in New York that he had "to-day hired a lawyer to drive a mule team. That is all the use lawyers are out here. We pay him $175 a month. Then, when you meet Judge White . . . tell him this." McGlynn's mother replied: "I saw Judge White and told him what you said, and he told me to say to you that he, as a lawyer, must say you could not have done better in the selection of a driver . . . for that the whole business of a lawyer is to know how to manage mules and asses, so as to make them pay."

The legal community emerged slowly in California. In large measure, the weak legal framework was born of indifference. Gold seekers were not interested in establishing a secure and orderly society; they were primarily motivated by greed. And they were impatient. As John David Borthwick wrote, "The every-day jog-trot of ordinary human existence was not a fast enough pace for Californians in their impetuous pursuit of wealth. The longest period of time ever thought of was a month." In this fast-paced atmosphere, building an enduring community featuring the usual social institutions was an afterthought. As historian Hubert Howe Bancroft concluded in 1888, they felt that "old maxims were as useless as broken crockery" in a world dedicated to impermanence and avarice. A traditional legal system would simply have to wait. Elisha Crosby, a forty-niner from New York who later became a leading California lawyer, politician, and civil servant, noted that "the proceedings of the California Courts in '49 . . . were rather uncertain. . . . In fact there was very little law of any kind, very few courts and very little proceedings."

A functioning legal system may have been of no interest to miners, but lawyers themselves were not immune to gold fever and, lured by the boom, some abandoned their practices back East in pursuit of the golden dream. These emigrants exhibited the same exuberance and desires as other gold seekers, and indeed many newly arrived lawyers had no intention of practicing law, instead ready to try their hands at mining. One diarist in Calaveras County remarked on the surprising vision of a lawyer in his courtroom attire panning for gold using not a pan but his stovepipe hat. But, like most hopeful miners, these attorneys had little success and eventually found themselves pursuing other employment. As mining camps and towns began to spring into existence around them and as the residents began to formulate rudimentary laws and regulations, many lawyers resumed legal practice. Most, however, did not forsake the impatience and unconventionality that had spurred them (and other Gold Rush emigrants) to pursue a new life in California in the first place.

In the mid-nineteenth century, the process of becoming a lawyer did not follow the path we are familiar with today. Lawyers seldom received formal legal training, and any white male meeting some general requirements could become certified. The standards were simple and straightforward: if you were at least twenty-one years old, were deemed to have good moral character, and had passed the test administered by a judge or practicing attorneys, you could officially open a law office. Most aspiring lawyers served an apprenticeship (usually called "reading the law") for a few months to prepare for the examination. Abraham Lincoln received a law license after reading borrowed legal books for less than a year; as he recalled years later, "I studied with nobody."

Some lawyers and jurists were sober and learned, while others were barely competent, or worse. There are numerous examples of outrageous decisions coming down through the California legal system

Hugh C. Murray, the youngest chief justice in the history of the California Supreme Court, took the position in 1852, when he was only twenty-six years old. In 1854 he wrote the notoriously anti-Chinese decision in *People v. Hall,* which barred Chinese people from testifying against white people in court. Courtesy of the California State Library, Sacramento; California History Section.

during this era. In Judge William Blackburn's court a habitual offender was found guilty of wife beating and sentenced to hang, but by law the trial proceedings of any death penalty cases had to be submitted, reviewed, and approved by the governor before execution could occur. Judge Blackburn was concerned that the dilapidated local jail, which had seen several escapes, could not hold the convicted felon overnight, so he ordered the criminal hanged immediately. The *next day,* Judge Blackburn sent the court papers for review by the governor. The judge scribbled a note on the papers indicating that, under the circumstances, it made no difference whether the execution should be approved or not.

Occasionally, these legal practitioners could be enormously entertaining and quirky, and the historical record abounds with anecdotes of shenanigans by less-than-stellar counselors at law and problematic jurists. Take lawyer Stanton Buckner, who was renowned for his long, windy speeches and summations. In one trial, Buckner kindly offered to explain to the judge the principle of innocent until proven guilty and launched into his disquisition. After several hours of listening to Buckner's interminable legal filibuster, the squirming judge interrupted the monotonous lawyer to announce, "The court is with you in that, but it does not admit that there is any presumption that the court's bottom is made of cast iron."

There are many tales of Francis J. Dunn, a lawyer known both for his theatrics and for his habit of imbibing a bit too much. One day, Dunn was found asleep in the road, and when he was aroused the tipsy fellow loudly proclaimed to all present, "I am Francis J. Dunn, considered, and justly considered, the best lawyer in the State of California." On another occasion, Dunn replied to an argument in court by saying, "The remarks of counsel remind me of a quotation from a classical poet. I cannot exactly recall the name of the poet, and I have forgotten the quotation; but, if I could repeat it, the court would see it is apropos." In another incident, Dunn was fined fifty dollars for contempt of court for arriving late. According to an account by Oscar T. Shuck:

> "I did not know I was late," said Dunn. "I have no watch, and I will never be able to get one if I have to pay the fines your Honor imposes upon me." (He had been fined before.) Then after a little reflection, Dunn said, "Will your Honor lend me $50 to pay this last fine?" "Mr. Clerk," said Judge Searls, "remit that fine. The State can afford to lose it better than I can." And the fine was remitted.

Some judges were just as idiosyncratic. During this period, cases were regularly decided without juries, and the judges not only made final pronouncements but also possessed the authority to dispatch cases as they saw fit. A colorful example was Judge William Almond, who was appointed as judge of the Court of First Instance (essentially a district court) by San Francisco's mayor John Geary. The byword in Almond's court was "swift justice." Jury trials were nonexistent, lawyers were limited to five-minute arguments, and most cases were settled within twenty minutes. The judge adopted a distinctive style as well. He dressed in his trail clothes and, according to Frank Soulé and James Nisbet in their 1855 *Annals of San Francisco*, "The judge sat upon a rickety old chair, with his feet perched higher than his head upon a small mantel over the fire-place, in which a few damp sticks of wood were keeping each other warm by the aid of a very limited supply of burning coals." While listening to cases, "His Honor employed himself in paring his corns, or scraping his nails." Judge Almond was contemptuous of lawyers prone to pomposity and flashes of pointless oratory, and he frequently stopped them in midsentence. Elisha Crosby observed the treatment of a lawyer named Lippitt, who was conducting his final summation when the judge interrupted: "You are a remarkably fine talker and you know a great deal of law, but the Court has not time to hear you now unless you insist upon it." Almond then announced his decision and the case concluded.

Judge Joshua Redman was another notable character. Elisha Crosby described Redman as "a man of very little education, rough and ready in his style and a hard drinker. He was utterly arbitrary in his decisions some of which were very peculiar, and were without regard to any known principle of law." In delivering judgements, Redman was fond of the phrase "If the Court knows itself, and the Court thinks it does," and multiple accounts recall that the judge habitually adjourned court with the words "If the Court knows itself, and the Court thinks it does, we will now adjourn the Court for five minutes and the Court will take a drink."

"ENJOY IT WITH LUXURIOUS ZEST"
JAMES MASON HUTCHINGS

He came seeking gold, but he found a jewel instead. Through his financial success as a miner, he was able to foster our endless admiration of an awe-inspiring natural garden. Without him hiring an itinerant mountain enthusiast, we might not have protected lands where we can always go to refresh our spirits. Without him, the most exquisite images of one of the most spectacular gorges on Earth may not have been created. But he was not a saint. At times he was consumed with envy and felt unappreciated. He could be stubborn, imperious, and covetous. This is the story of that man and the "singular and romantic valley" he loved: Yosemite Valley.

The life of James Mason Hutchings began in England in 1820, but he was reborn during the Gold Rush. He immigrated to the United States in 1848 and settled in New Orleans to pursue a promising business opportunity. But his residency in the Crescent City was short-lived, as he lit out for California in 1849 after contracting a case of gold fever. In October 1849, he arrived in the land of golden possibilities and purchased a "mining hole" near Placerville, as he recollected in his diary. Hutchings calculated that he had traveled 5,106 miles from England to El Dorado County.

In a letter published in the *New Orleans Daily Picayune* on December 19, 1849, he recalled his earliest days as a miner. Hutchings realized $5.75 his first day in the placers with a pan, $10.60 the second day using a borrowed cradle, and $27.30 his third, for a total of $43.50. He had had better luck than many miners after three days' labor, earning a sum that, he noted, "in New Orleans would have taken me two weeks."

Working both alone and with a consortium of miners in the communities of Georgetown, Coloma, and White Rock Spring, and in the area near Placerville, Hutchings was lucky and garnered several thousand dollars in profit. He deposited the earnings in a San Francisco bank, the Frank Ward and William Smith Mercantile and Auction Company, with the intention of purchasing several city lots in the booming City by the Bay at an auction scheduled for February 22, 1850. Just a few days before the auction, however, the bank collapsed. Hutchings lost everything.

He could have thrown in the towel, but he persevered. He returned to mining at Michigan Bluff, near Auburn, Placer County. Once again luck was with him, and he purchased Rock Spring Ranch, outside of Auburn, in 1850. Hutchings then traveled to the rich mining area known as Weber Creek in El Dorado County, a claim established by Charles Maria Weber, the founder of Stockton. Hutchings struck it rich and cleared $8,000 within a year.

It was during this period that Hutchings accidentally acquired a new career path.

In the goldfields, Sunday was the principal business day. Critics felt that the day should be reserved for religious observances, but merchants argued that the Sabbath was already firmly established as market day and that any attempt at proselytizing for change would be scorned by the miners. James Hutchings penned a tongue-in-cheek observation in the form of a "commandment," which read in part:

> Thou shalt not remember what thy friends do at home on the Sabbath day, lest the remembrance may not compare favorably with what thou doest here.

The pronouncement was immediately popular, and Hutchings expanded his one commandment to ten, publishing "The Miners' Ten Commandments" in 1853. Printed as a letter sheet, it sold nearly one hundred thousand copies. Hutchings earned enough money that he could quit mining and pursue a new venture.

In 1855, he became fascinated with the Mariposa Battalion's account of the Yosemite Valley, which they had "discovered" in 1851 during the Mariposa Indian War of 1850–51. The war had sprung from the conflict between Native Sierrans and the encroaching gold seekers in the area just west of present-day Yosemite National Park. As the region was overrun with argonauts who stole their land, the Indians grew resentful and increasingly angry. An Indian uprising was rumored, but few believed it possible until a series of raids in December 1850 near Fresno Crossing destroyed a trading post and killed three white employees. With the local settlers out for blood, California's governor, John McDougall, was pressured into authorizing the establishment of the two-hundred-member "Mariposa Battalion" to combat the threat. The battalion waited while a Federal Indian Commission sought a peaceful resolution, and in March 1851 a peace treaty was ratified with six local tribes. However, members of the Yosemite Miwok, Chowchilla, and Yokuts tribes were not present at the signing, and military action was commenced against them immediately. Three campaigns were launched, and forty people were killed in bloody skirmishes before the Native resistance ended in May 1851. The most famous campaign was the first foray against the Yosemite tribe, which had led the battalion into Yosemite Valley in late March.

The narrative of the campaigns was thrilling to Hutchings, particularly the wonderful descriptions of the extraordinary valley. He resolved to see for himself if there truly was a waterfall "nearly a thousand feet high," as battalion members had reported. He hired two Native guides to take him to the site and, in spring of 1855, James Mason Hutchings first beheld the "singular and romantic valley" that was to change his future. Entrepreneur to his marrow, Hutchings instantly appreciated the economic potential of the extraordinary landscape, and he addressed this vision in the *Mariposa Gazette* of August 9, 1855:

> After completing our series of views of this beautiful and wildly romantic valley, we looked a last look upon it, with regret that so fine a scene should be only the abode of wild

> animals and Indians, and that many months, perhaps years, would elapse before its silence would again be broken by the reverberating echoes of the rifle, or the musical notes of the white man's song. . . . I have no doubt ere many years have elapsed, this wonderful Valley will attract lovers of the beautiful from all parts of the world; and can be as famed as Niagara, for its wild sublimity and magnificent scenery.

Noting the popularity of his letter sheet and the public's desire for more information about Yosemite Valley in particular, Hutchings determined that Gold Rush California required a monthly periodical. After more than two years and in excess of $6,000 toward its development, *Hutchings' Illustrated California Magazine* emerged from its cocoon in 1856. The inaugural edition featured the first published illustrations of Yosemite, and overnight they turned the valley into a captivating tourist phenomenon. Hutchings urged his readers to visit the wonderland that awaited them in the Sierra Nevada and "enjoy it with luxurious zest." Ultimately, millions would take his advice.

Hutchings' Illustrated California Magazine published sixty liberally illustrated issues between 1856 and 1861. James Mason Hutchings was publisher, editor, and ringmaster. The endless task of producing the elaborate magazine was, unfortunately, physically and emotionally stressful and seriously impacted Hutchings's well-being. As explained in a reminiscence of Hutchings released in 1901 by the Society of California Pioneers, publishing was "undermining his health, and . . . he must select between two alternatives, either to give up his profession or leave this world." In the end, "he chose the former as being the easiest, most acceptable and the best comprehended."

In 1864, resolutely convinced of the remarkable commercial promise of Yosemite, Hutchings bought the only hotel in the valley, the ramshackle Upper Hotel, which was renamed Hutchings House. The hotel accommodations were primitive, but Hutchings made some improvements, including installing glass in the open-framed windows, providing fern-stuffed mattresses, and creating "rooms" by hanging sheets as partitions. The hotel's most interesting feature was constructed in 1866: a parlor, known far and wide as the Big Tree

Room, fashioned around a large incense cedar. In his 1894 *Souvenir of California*, Hutchings wrote:

> This cedar, 175 feet high, was standing there when the room was planned. I had not the heart to cut it down, so I fenced it in, or rather, built around it. . . . The base of the tree, eight feet in diameter, is an ever present guest in that sitting room. . . . The large, open fireplace was built with my own hands. . . . Travelers from all climes and countries welcomed the sheltering comfort and blazing log fire of this room.

In 1869, Hutchings hired a thirty-one-year-old itinerant Scottish naturalist to build and manage his sawmill in the valley. His name was John Muir. Muir had yearned to visit Yosemite for years and booked passage to San Francisco from New York via Panama in 1868. He reached San Francisco on March 27, 1868, and when he was asked, "Where are you going?" upon first setting foot in the city, he reportedly replied: "Anyplace wild." Muir walked 165 miles to Yosemite Valley, arriving at the majestic venue on May 22, 1868. After investigating the valley environs, Muir found employment with Hutchings. Muir lived at the sawmill he oversaw in an upstairs dwelling he dubbed the "Hang Nest."

Friction grew between the two men as the energetic and charismatic Muir quickly gained a reputation as the foremost Yosemite naturalist and spokesman. Hutchings felt that he had an earlier and greater claim to that position and resented Muir's emerging fame. Muir quit after two years. It is telling that when Hutchings published his 1888 memoir of his life in Yosemite, *In the Heart of the Sierras*, John Muir is not mentioned by name in the entire 496-page book.

Only two months after Hutchings purchased the hotel, President Abraham Lincoln signed the Yosemite Valley Grant Act of 1864—legislation that brought the Yosemite Valley and Mariposa Grove of Big Trees under government control. While this should have been good news for Hutchings, it was not. The act left all private ownership claims in legal limbo, and Hutchings was informed that his title to Hutchings House was invalid. He would have to sign a lease to continue operating his business. Hutchings disagreed, and a ten-year legal battle ensued. In 1874, Hutchings received a $24,000 settlement from the California legislature. He had spent more than $41,000 during the litigation. Hutchings sold the hotel and moved to San Francisco.

James Mason Hutchings returned to Yosemite in 1880 when the commissioners of the Yosemite Grant named him Guardian of the Mariposa Grove and Yosemite Valley, replacing Galen Clark, who had held the post for more than twenty years. Hutchings's supporters felt his tenure was characterized, as the Society of California Pioneers put it, by "remarkable acceptability and efficiency for several years." His detractors found him prickly, gauche, and overbearing. Hutchings was discharged after four years and left Yosemite.

James Mason Hutchings and his wife, Emily, on October 31, 1902, the day he died. Photograph from *James Mason Hutchings of Yo Semite,* by Dennis Kruska (San Francisco: Book Club of California, 2009). Kruska Collection, Figure 88. Courtesy of Dennis Kruska.

James Mason Hutchings and his family did return occasionally to the valley for visits and holidays. On the crisp late afternoon of October 31, 1902, the white-haired eighty-two-year-old, wearing a stylish suit and sporting a dapper chapeau, was on a Halloween carriage ride with his third wife, Emily, heading east to the spectacular Sierra Nevada cathedral of granite. As they descended Big Oak Flat Road toward Yosemite Valley, the elderly man pulled the reins and stopped when the monolith of El Capitan came into view. Exhilarated, he stood up in the carriage and loudly exclaimed, "It looks like Heaven!"

Hutchings and Emily drank in the view for an instant. Perhaps he remembered the summer day forty-seven years earlier when he first spied the Yosemite Valley. He had described that moment in the August 9, 1855, edition of the *Mariposa Gazette*:

> We climbed nearly to the ridge of the middle or main fork of the Merced, and then descending towards the Yo-Semity Valley, we came upon a high point, clear of trees, from where we had our first view of this singular and romantic valley; and as the scene opened in full view before us, we were almost speechless with wondering admiration, at its wild and sublime grandeur.

Lost in their reverie, the couple did not notice a wild animal creeping near their carriage. It spooked their horses and, in a twinkling, the moment and the Hutchings's world dissolved into chaos.

As Hutchings stood in the carriage extolling the virtue and beauty of El Capitan, his horses bolted. Emily was thrown clear of the vehicle as it careened wildly down the road. Twenty feet later, Hutchings was himself violently hurled from the vehicle. He bounced and somersaulted to an abrupt stop when his head struck a jagged pile of rocks. Emily rushed to his side. She cradled her bloodied husband in her arms as Hutchings whispered his last words: "I am very much hurt." As Emily wrote in the register of the Sentinel Hotel on November 8, 1902, "The Angel of Death reached Mr. Hutchings a few moments later." The remains of James Mason Hutchings were carried into the valley and his funeral service was held in the Big Tree Room of his old hotel. He is buried in the Yosemite Valley Pioneer Cemetery, in the shadow of Yosemite Falls. His rough granite tombstone reads "Father of Yo Semite."

Years later, Hutchings would have still another influence on the history of Yosemite. In 1916, a fourteen-year-old boy who was sick in bed in San Francisco was given a book to read: the Hutchings memoir *In the Heart of the Sierras.* The boy was enthralled by the story and begged his parents to vacation in Yosemite National Park when he recovered. That summer, they traveled to the park. The boy was equipped with a present from his parents, a Kodak Brownie camera. He gleefully tramped through the park and documented his memorable visit with many snapshots. He fell in love with the valley that summer, and the love affair never faded. His name was Ansel Adams, the preeminent photographic chronicler of Yosemite.

LETTER OF THE LAW

LETTER SHEETS AND THE TEN COMMANDMENTS

James Mason Hutchings was an indefatigable promoter of the Golden State. His brainchild and promotional tool was his *Hutchings' Illustrated California Magazine*, which was published from 1856 to 1861 with a circulation in excess of eight thousand. But it would not have been possible without the sale of one pale-blue piece of paper measuring eleven by eighteen inches.

The paper was a pictorial letter sheet, a form of Gold Rush stationery featuring engraved illustrations that could be folded into an envelope and mailed for the minimum postal fee of forty cents. A simple, relatively inexpensive method for relaying news back home, the letter sheet was a familiar Gold Rush item, and hundreds of thousands were printed featuring more than three hundred scenes of mining operations, spectacular landscapes, daily life in the gold diggings, and bird's-eye views of large and small communities available for purchase. The most popular of all the letter sheets was called *The Miners' Ten Commandments*. It was the product of the prolific pen of James Hutchings and the clever drawings of Harrison Eastman. More than one hundred thousand copies of this letter sheet were sold.

Hutchings's *Commandments* were a hit, but using that specific Biblical framework was not a new concept. The idea that morality was completely set aside during the Gold Rush has been much exaggerated, but in fact religion and spiritual guidance remained respected components of the culture, albeit with some customization within the often rough-and-tumble mining camps. That said, religious conviction—or lack thereof—was a common target of critical observation, and false piety in particular was ripe for satire. In March 1849, the *Cleveland Plain Dealer* published "The Wife's Twelve Commandments" for husbands, which included the following mandates:

> Thou shalt have no other wife but me.
>
> Thou shalt not chew tobacco.
>
> Thou shalt not covet the tavern keeper's rum, nor his brandy, nor his gin, nor his whisky, nor his wine, nor anything that's behind the bar of the rum-seller.
>
> Thou shalt not stay out later than nine o'clock at night.

In 1853, Hutchings wrote *The Miners' Ten Commandments* for the *Placerville Herald*. Typical of the usually short-lived Gold Rush newspapers, the *Herald* was only published for a few months in 1853 and then expired. Even so, Hutchings's commandments were the most well-liked feature the newspaper ever printed.

Hutchings began by referencing a fashionable phrase of the era, "going to see the elephant," which was a slang term for setting off on an adventure:

> I am a miner, wandering "from away down east," to sojourn in a strange land. And behold I've seen the elephant, yea, verily, I saw him, and bear witness, that from the key of his trunk to the end of his tail, his whole body hath passed before me; and I followed him until his huge feet stood before a clapboard shanty; then with his trunk extended he pointed to a candle-card tacked upon a shingle, as though he would say Read, and I read the
>
> MINERS' TEN COMMANDMENTS

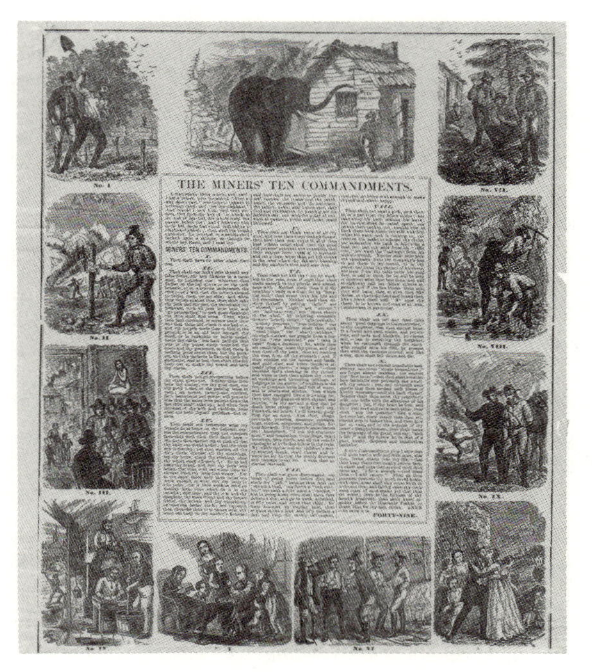

THE MINERS' TEN COMMANDMENTS.

The most popular letter sheet of the Gold Rush was the fanciful *Miners' Ten Commandments,* 1853. Courtesy of the California State Library, Sacramento; California History Section.

The Miners' Ten Commandments mixed humorous content with serious admonishments for mostly young, single miners. It advised not to make false claims or steal picks, shovels, and pans, nor to claim jump, nor to dishonor your parents with questionable behavior. It urged the argonauts not to grow discouraged, and to remain faithful to their values and to those they had left behind, such as "thy first love." It reminded the married miners that failure is not unusual and that it is appropriate to let your family know when you are heading back home.

But what most captivated readers were the witty asides and silly allusions—some of which ran hundreds of words in length—that reflected the realities of the mining camps. In addition to the more extensive thoughts couched in faux religiosity, Hutchings intermingled these punchy, light-hearted phrases:

> Thou shalt have no other claim than one.
>
> Thou shalt not go prospecting before thy claim gives out. Neither shalt thou take thy money, nor thy gold dust, nor thy good name, to the gaming table in vain . . . for [it] will prove to thee that the more thou puttest down the less thou shalt take up.
>
> Thou shalt not remember what thy friends do at home on the Sabbath day, lest the remembrance may not compare favorably with what thou doest here.
>
> Thou shalt not kill. . . . Neither shalt thou destroy thyself by getting "tight," nor "stewed," nor "high," nor "corned," nor "half-seas over," nor "three sheets in the wind," by drinking smoothly down—"brandy slings," "gin cocktails," "whiskey punches," "rum toddies," nor "egg nogs." Neither shalt thou suck "mint juleps," nor "sherry-cobblers," through a straw.
>
> Thou shalt not grow discouraged, nor think of going home before thou hast made thy "pile."
>
> Thou shalt not tell any false tales about "good diggings in the mountains."

The letter sheet was illustrated by the drawings of Harrison Eastman, a forty-niner from New Hampshire. Eastman was not a miner but a self-taught artist, and he quickly found employment as an engraver and lithographer. After a few months in San Francisco, he traveled to Sacramento and resumed work as a sought-after wood engraver. He designed the masthead for the *Sacramento Transcript* newspaper as well as the official seal of Sacramento City. Eastman stayed in Sacramento less than a year and returned to San Francisco, where he established a thriving printing and engraving business, and in 1853, he was engaged by Hutchings to illustrate the letter sheet for *The Miners' Ten Commandments*. His drawings included a large image of an elephant pointing at the list nailed on a cabin shingle plus ten smaller, more literal illustrations of the individual decrees. Eastman later became a primary illustrator for *Hutchings' Illustrated California Magazine*.

The popularity of the Hutchings/Eastman letter sheet spawned a host of imitators, and many versions of the commandments were circulated over the years. The best known was an 1855 version, published by Hutchings and illustrated by W. C. Butler, titled *Commandments to California Wives.* Following the template established by Hutchings, these commandments begin with a lonely miner approached in his cabin by an angel "clothed in female apparel." The angel grabs him by "the bosom of [his] woolen shirt" and, "in a voice of musical distinctness," asks if he is sufficiently courageous to hear her words. The trembling miner modestly answers that he had little courage of which to boast, but he is trying. "It is enough," the angel replies, and then delivers unto him the *Commandments to California Wives.* Like its predecessor, this spinoff sprinkled wry (albeit strikingly sexist) commentary among the warnings to women not to dishonor their husbands, or waste money, or complain or gossip excessively. The author urges wives to be encouraging and thoughtful to their menfolk even when they do not really deserve it. He reminds women that the life will be difficult, that delicate flowers will wilt, and that they should expect very little pampering, and he doles out these directives:

> Thou shalt not "put on airs" of self-importance, nor indulge in day-dreams of extravagance . . . or make thee MERELY an expensive toy and walking advertisement of the latest fashions.
>
> Thou shalt not believe thyself to be an angel—all but the wings.
>
> Thou shalt not substitute sour looks for pickles. . . . Neither shalt thou serve up cold looks or cold meats for breakfast, nor scoldings and hard potatoes for dinner.
>
> To Unmarried Ladies. Thou shalt not . . . marry . . . because he is rich—(for here the rich become poor and the poor become rich.)

"TO THE LAND OF GOLD AND WICKEDNESS"

LORENA LENITY HAYS BOWMER

In April 1850, forty-niner Joseph Crackbon was in Nevada City, frustrated, homesick, and lonely. He had left his home in Massachusetts more than a year before, and his journey to California across the Isthmus of Panama had been arduous. Once in the diggings, Crackbon's forays into gold panning were unsuccessful. He was woefully disheartened. And then a vision appeared. As he wrote in a letter that day: "Got nearer to a woman this evening than I have been in six months. Came near fainting."

In October 1849, gold seeker William Prince, then in Stockton, wrote to his wife that there were so "few females that I never yet have seen one in the streets." On September 20, 1851, Louise Amelia Knapp Smith Clappe, better known as Dame Shirley, wrote about a young founder of Rich Bar, a mining hamlet on the Feather River, "This unfortunate had not spoken to a woman for two years, and, in the elation of his heart at the joyful event, he rushed out and invested capital in some excellent champagne, which I . . . assisted the company in drinking, to the honor of my own arrival." The following year Dame Shirley described herself as a "petticoated astonishment" to the men in nearby Indian Bar.

Women were certainly scarce during the Gold Rush, but the exact numbers are difficult to pin down. The rapid influx of new Californians, and their tendency to move frequently, made the results of the state's 1850 census unreliable at best. However, the numbers that are available indicate that women made up about 8 percent of California's population in 1850. In the goldfields, the number hovered around 3 percent, although it was higher in larger Gold Country communities and lower in the more remote outposts. Some recent scholarship suggests that the percentage of women was a bit higher overall, as some types of women, such as prostitutes, were not normally recorded.

Perhaps based on their tiny numbers, for more than a century women were considered to be merely marginal participants in the Gold Rush. In fact, many books do not even list a single woman in their indexes. That notion was brilliantly dispelled by the groundbreaking work of JoAnn Levy in her 1992 book *They Saw the Elephant: Women in the California Gold Rush.* Levy documents the impact of women on the event and debunks the myths saying either that the Gold Rush was exclusively male or that if women were present, their activities and contributions were negligible.

The truth is that women, through their letters, diaries, and journals, provide an invaluable resource to understanding the Gold Rush. Many women who traveled to the goldfields were well-educated and possessed extraordinary facility with the written word, and some were published authors. A handful of these writers remain well known, such as Dame Shirley and Luzena Stanley Wilson, but most are denizens of the historical outskirts, relegated to the shadowy realm of nearly forgotten history. This is unfortunate because their stories open fascinating windows into life during this tumultuous period.

The Gold Rush was front-page news in American cities both large and small as people across the land clamored for specifics from the goldfields. Dispatches from eyewitnesses were especially desirable, and since the vast majority of the participants and writers were male, any correspondence providing a woman's viewpoint was seized upon. Descriptions by women of everyday experiences, whether sensational or mundane, helped satisfy this growing thirst for knowledge. These writings also usually offered a more intimate glimpse of the mining camps and commercial centers, and they were welcome windows into daily life for those women and girls separated from their husbands or beaus; they appreciated any detail that could help span the physical and emotional distance. While the women of the Gold Rush observed the same cultural landscape as the men, often they did so with different sensibilities and priorities, and therefore their perspectives helped to provide a much fuller, more well-rounded vision of the event. These narratives are an invaluable piece of the Gold Rush mosaic. For example, consider the case of Lorena Lenity Hays Bowmer.

Born in Pennsylvania in 1828, Lorena was the oldest child in a family of nine brothers and sisters that eventually settled in Illinois. She was a hypochondriac who always remained hopeful that her health would improve, perhaps tomorrow. In 1848, Lorena's father died and the family suffered immediate financial hardship. In 1852, her Uncle Henry came back to Illinois after a brief visit to California, bringing with him astonishing accounts of the golden land. Henry had a hankering to return to California as soon as possible, and he persuaded his family to join him in seeking a better life in the faraway El Dorado. In March 1853, Lorena, Uncle Henry, and other members of her family embarked on a six-month wagon train journey to California as part of the Woodward-Hays-Lithgow Company. Lorena kept a diary, which was later published as *To the Land of Gold and Wickedness.*

Lorena was prepared for a difficult road. "I expect to see some hard times," she wrote in her diary. "Oh! may I be firm and try to set a good example." She walked nearly the entire route, only intermittently riding on horseback. The party supplemented their diet with buffalo meat, and they frequently cooked over buffalo chips gathered by her younger sisters. Lorena rhapsodized about the blankets of perfumed wildflowers but bemoaned the choking clouds of alkali dust and buzzing swarms of mosquitoes and blood-sucking "buffalo gnats" (also known as black flies.)

In addition to detailing the fine points of the crossing, Lorena's observations from the trail provide unique insights into the attitudes and concerns of a young woman traveling to California. She writes of employment opportunities out West, "There seems to be a good prospect for females to make good wages there." She compares the advantages and disadvantages of migration to California on the outlook for

Pennsylvanian Lorena Lenity Hays Bowmer was one of the many thousands of confident migrants that traveled the winding, challenging path to the Land of Gold. "Emigrant Party on the Road to California," frontispiece from *California: Its Past History, Its Present Position, Its Future Prospects* (1850). Courtesy of the Library of Congress, Washington, D.C.; ID: LC-DIG-ppmsca-02887.

marriage, and mentions a deception practiced by some men on the overland trek: "Several gentlemen who are married have been passing themselves off as unmarried."

By July, the party reached South Pass in Wyoming, the lowest-elevation pathway through the Rockies on the Oregon-California Trail. It was midsummer, but the snow and ice at South Pass was fifteen feet deep. A few days later, farther west, the sojourners began traversing the blazing Forty Mile Desert. Near Lorena's company were other wagon trains headed to California, and she noted more than one thousand wagons and three hundred thousand head of cattle. The days became monotonous, water was scarce, shade was infrequent, the smell of dead oxen rotting beside the road permeated the stifling air, and the earlier poetic tone of her diary disappeared as the entries became terse.

By the end of August 1853, the wagon train had crossed the desiccated barrens and the Sierra Nevada came into view. Tempers grew short as the journey neared its end, and in early September, members of the wagon company murdered a fellow traveler over the ownership of a steer. Lorena recorded the incident succinctly: "It seems the father and son who shot the young man had but little provocation. The son made his escape." A trial was held a few weeks later, but the killers were freed. There were no witnesses present and no one appeared to testify against the murderers. They had all gone to California.

Lorena's heart leapt with excitement as they slowly crept toward the snowy range of the Sierra. On September 10, 1853, from the Carson Valley, she wrote: "The great Sierra Nevada towers up in front of us over which we shall soon have a toil, and from whose summit we shall take a view of the far-famed gold land, the land of our hopes. O' Shall they be realized? Shall we be happier in our new home?" The wagon party edged closer to the end of this dusty chapter as they headed toward residence in Amador County, near the town of Ione.

Three days later, as her company started to climb the Carson River Canyon at the foot of the eastern Sierra Nevada, Lorena exalted in the atmosphere:

> We breathe the exhilarating mountain air, look at the towering pines and lofty mountain peaks, and listen to the music of the dashing waters, as they fitfully leap, and foam over their rocky beds, and hear the distant sighing and murmuring, and the now nearer loud roaring of the wind as it comes sweeping onward the pine trees, and we scramble over rocks and descend into quiet vales.

Lorena Lenity Hays arrived in Amador County on September 29, 1853. She recalled,

> I am no longer to live a camp life, no longer to walk, ride and to be almost constantly in the fresh, free air of Heaven. A roof must now intervene between me and the blue dome above. . . . I can never feel so free, so independent anymore. . . . The last part of the journey was rough, and somewhat tedious, but I have no reason to regret it in the least.

It wasn't long before Lorena was keenly aware that this "far-famed gold land" was a new world with novel challenges. She soon realized that the society she was entering was unruly, uncivilized, unusual, unfamiliar.

> This is a wild, romantic and almost barbarous country. It seems as if all restraint were laid aside by every one when they arrive here. Kind admonitions, earnest prayers and good advice of pious, anxious parents at home are quite forgotten, and sober-minded young men of good habits become wild and reckless. . . . Now I am situated in a strange land amid strangers.

The family struggled. They took in boarders, including a starving and destitute miner who soon died. She wrote, "How hard it must be for one so young, and beautiful, to die far away from home in this land of strangers."

On February 17, 1855, Lorena married John Clement Bowmer. The marriage ceremony was a bit of a surprise, Lorena cryptically remarked, as it occurred "rather unexpectedly in church, owing to some misunderstanding between self-and-*husband*." Lorena never explained how she accidentally got married, but she was candid in her feelings toward her new husband. Lorena did not love Clement, as she called him, "but *he* is good, and kind, and devoted." Clement eventually presented Lorena with a wedding ring, which featured twelve miniature diamond chips.

Lorena had a daughter in 1856 who died soon after birth. A son named Henry came later. He grew up to become a prominent newspaper publisher and reporter in Nevada and Washington, either writing for, founding, purchasing, or managing twenty-two newspapers over the next fifty-seven years. He must have inherited the writer's gene from his mother, who had begun to publish her reflections in *The Golden Era*, a well-known San Francisco literary newspaper, not long after marrying Clement. As was the custom of most authors during that period, she used pseudonyms, writing as "Lenita" or "Our Mountain Lassie."

Mostly Lorena's comments were cheerful, but occasionally her thoughts were serious meditations on goldfield society. In an April 8,

1855, letter printed on the front page of *The Golden Era*, she wrote:

> The [gambling and drinking] establishments here have sent forth a miasma that has polluted the moral atmosphere of all our towns. . . . But . . . I am forgetting that it is news you most desire in a letter and that you prefer the "sunny side" of a story to dull dissertations on morality or other every-day-written-upon subjects.

Deeply religious, Lorena urged her female readers to summon their "secret power" from God to lead needed reforms in California. She predicted that the outcome of this moral campaign would be that "the gambler, duelist, Sabbath-breaker and drunkard would soon cease to stalk in insolent dignity through the land."

An October 1855 offering entitled "A Chapter on Beards" offered a lighter musing on Gold Rush culture:

> There is something in a finely adjusted head of hair and properly cultivated beard, that most women cannot help but admire. Perhaps it imparts nobleness and dignity, and a look of strength. . . . Then let him wear the emblem of his strength. But let him select a becoming style, for one sees such outlandish faces now-a-days to disgust him with all bearddom. Some faces resemble ferocious tigers, others silly goats or senseless buffoons. We think a beard-less phiz. preferable to these burlesques upon mankind.

Sadly, Lorena Lenity Hays Bowmer did not live to see many more of her comments published. She fell ill, probably from typhoid fever, and on October 30, 1860, she died at the age of thirty-two. She is buried somewhere in Ione; the location of her grave is unknown.

The Bowmer legacy lived on, however, and continued to be a literary one. Lorena's great-grandson, Angus Bowmer, founded the world-renowned Oregon Shakespeare Festival in Ashland in 1955. When he passed away in 1979, Angus had in his possession a piece of jewelry he had inherited from his father. It was a ring containing two of the tiny diamond chips from Lorena's wedding band.

"THE THEATRE OF UNREST"
ELIZA FARNHAM

This indomitable spirit was a prison reformer, author, lecturer, abolitionist, early feminist, and, most of all, a force of nature. She could rage at the storm, confront the devil, and triumph over insult, violence, and tragedy, but also plant potatoes, wield a hammer, and tenderly nurse wounded warriors. She is infrequently recognized today, but during the California Gold Rush, the name of Mrs. Eliza Wood Burhans Farnham was well known—especially for one incident in 1849.

Thomas Jefferson Farnham, Eliza's husband, was in California practicing law, writing, and tending to a two-thousand-acre spread near Santa Cruz in 1848. Eliza and her children, then in Boston, hoped to soon join Thomas in the distant province.

And then their lives dramatically changed.

Thomas died on September 13, 1848, of "intermittent fever," a malarial disorder. Eliza made immediate plans to travel to California to reconcile his affairs, but she also wanted the journey to have meaning and social consequence beyond simply a voyage of bereavement and estate settlement. After all, she would write later, "California will not always be the theatre of unrest, of reckless hazard, and unscrupulous speculation," and perhaps she could facilitate its advancement.

In February 1849, Eliza Farnham announced her plans for a "California Association of American Women" in a widely distributed broadside. The first sentences read:

> The death of my husband, Thomas J. Farnham, Esq., at San Francisco, in September last, renders it expedient that

> I should visit California during the coming season. Having a desire to accomplish some greater good by my journey thither than to give the necessary attention to my private affairs, and believing that the presence of women would be one of the surest checks upon many of the evils that are apprehended there, I desire to ask attention to the following sketch of a plan for organizing a party of such persons to emigrate to that country.
>
> Among the many privations and deteriorating influences to which the thousands who are flocking thither will be subjected, one of the greatest is the absence of woman, with all her kindly cares and powers, so peculiarly conservative to man under such circumstances.

In the remainder of the lengthy circular, Eliza detailed that an assembly composed only of unencumbered females would travel with her to California and offer the "benefits that would flow to the growing population of that wonderful region, from the introduction among them of intelligent, virtuous, and efficient women."

The company would be women "not under twenty-five years of age," and sanctioned by their local clergyman or town authority, who could "contribute the sum of two hundred and fifty dollars, to defray the expenses of the voyage [and] make suitable provision for their accommodation after reaching San Francisco." She predicted that a group of "one hundred or one hundred and thirty persons" with "six or eight respectable married men and their families" would be easy to secure.

Eliza Farnham's proposal was a sensation for its truthfulness and startling boldness. The indisputable reality was that women were rare in California and much desired by the lonely gold seekers for companionship. Additionally, many argued that the young, unattached men who formed the bulk of the argonaut population were in danger of slipping into chaos without women's "civilizing" influence. Peter Burnett, an emigrant to the goldfields who would later become California' first governor, noted that "in California, . . . there were few women and children, but plenty of gold, liquors, and merchandise, and almost every man grew comparatively rich for the time; and

SHIP ANGELIQUE.

CALIFORNIA ASSOCIATION OF AMERICAN WOMEN.

NEW YORK, FEBRUARY 2D, 1849.

THE death of my husband, THOMAS J. FARNHAM, Esq., at San Francisco, in September last, renders it expedient that I should visit California during the coming season. Having a desire to accomplish some greater good by my journey thither than to give the necessary attention to my private affairs, and believing that the presence of women would be one of the surest checks upon many of the evils that are apprehended there, I desire to ask attention to the following sketch of a plan for organizing a party of such persons to emigrate to that country.

Among the many privations and deteriorating influences to which the thousands who are flocking thither will be subjected, one of the greatest is the absence of woman, with all her kindly cares and powers, so peculiarly conservative to man under such circumstances.

It would exceed the limits of this circular to hint at the benefits that would flow to the growing population of that wonderful region, from the introduction among them of intelligent, virtuous and efficient women. Of such only, it is proposed to make up this company. It is believed that there are hundreds, if not thousands, of such females in our country who are not bound by any tie that would hold them here, who might, by going thither, have the satisfaction of employing themselves greatly to the benefit and advantage of those who are there, and at the same time of serving their own interest more effectually than by following any employment that offers to them here.

It is proposed that the company shall consist of persons not under twenty-five years of age, who shall bring from their clergyman, or some authority of the town where they reside, satisfactory testimonials of education, character, capacity, &c., and who can contribute the sum of two hundred and fifty dollars, to defray the expenses of the voyage, make suitable provision for their accommodation after reaching San Francisco, until they shall be able to enter upon some occupation for their support, and create a fund to be held in reserve for the relief of any who may be ill, or otherwise need aid before they are able to provide for themselves.

It is believed that such an arrangement, with one hundred or one hundred and thirty persons, would enable the company to purchase or charter a vessel, and fit it up with every thing necessary to comfort on the voyage, and that the combination of all for the support of each, would give such security, both as to health, person and character, as would remove all reasonable hesitation from the minds of those who may be disposed and able to join such a mission. It is intended that the party shall include six or eight respectable married men and their families.

Those who desire further information will receive it by calling on the subscriber at

ELIZA W. FARNHAM.

The New-York built Packet Ship ANGELIQUE has been engaged to take out this Association. She is a spacious vessel, fitted up with state rooms throughout and berths of good size, well ventilated and provided in every way to secure a safe, speedy and comfortable voyage. She will be ready to sail from New-York about the 12th or 15th of April

---

WE, the undersigned, having been made acquainted with the plan proposed by MRS. FARNHAM, in the above circular, hereby express our approbation of the same, and recommend her to those who may be disposed to unite with her in it, as worthy the trust and confidence necessary to its successful conduct.

HON. J. W. EDMONDS, Judge Superior Court.
HON. W. T. McCOUN, Late Vice Chancellor.
HON. B. F. BUTLER, Late U. S. Attorney.
HON. H. GREELEY.
ISAAC T. HOPPER, ESQ.
FREEMAN HUNT, ESQ.
THOMAS C. DOREMUS, ESQ.

W. C. BRYANT, ESQ.
SHEPHERD KNAPP, ESQ.
REV. GEORGE POTTS, D. D.
REV. HENRY WARD BEECHER.
MISS CATHARINE M. SEDGWICK.
MRS. C. M. KIRKLAND.

NESBITT, PRINTER.

In February 1849, New York social reformer Eliza Farnham initiated a novel and audacious plan to assemble a company of unmarried women to travel to Gold Rush California under the name California Association of American Women. California Historical Society, San Francisco, Vault-B-004.jpg.

yet, in the absence of female influence and religion, the men were rapidly going back to barbarism." This sentiment was echoed in October 1849, when the *Alta California* published a passage from a forthcoming book on California's prospects by Dr. Felix Wierzbicki, a Polish immigrant and veteran of the Mexican War:

> We will try to advocate the cause of poor and forlorn bachelors, and persuade some respectable heads of families that have daughters to settle in life, to come to California and build up the society, which, without woman, is like an edifice built on sand. Woman, to society, is like a cement to the building of stone; the society here has no such a cement; its elements float to and fro on the excited, turbulent, hurried life of California immigrants, or rather gold hunters, of all colors and shapes, without any affinity. . . . But bring woman here, and at once the process of crystallization . . . will set in in the society, by the natural affinities of the human heart.

Nothing similar to Farnham's plan had ever been proposed. Daily, newspapers east of the Mississippi listed dozens of ships and wagon companies headed to California, but overwhelmingly the passengers were male. The idea of transporting one hundred or more women, all at once, to California was innovative and daring. Some predicted Eliza's California Association of American Women would suffer failure and personal embarrassment. But the expedition was quickly endorsed by prominent community leaders, including author Catharine Sedgwick, newspaper publisher Horace Greeley, poet and journalist William Cullen Bryant, and abolitionist firebrand Henry Ward Beecher. Their confidence stemmed from cognizance of Eliza Farnham's personal fortitude and extraordinary life story.

Born in New York in 1815, Eliza Wood Burhans had a challenging childhood. Her mother died when she was five years old, and Eliza was placed with foster parents. She had contempt for her actual father, who, she believed, cheated on her mother before her death. Her foster mother was brutally authoritarian, her foster father weak and unsupportive, and Eliza was physically and emotionally abused. As an adult, Thomas, her first husband, was often absent, and two of her three children died in childhood. She suffered through a horrifying second marriage to a violent alcoholic. Understandably, with her world filled with abuse and uncertainty, Eliza welcomed the possibility of change in her personal life and came to embrace reform as an indispensable solution to societal cruelties as well.

In July 1836, she married Thomas Jefferson Farnham, a peripatetic young Illinois lawyer subsequently noted for his travel writings about the American West, for his advocacy of settlement in California, and for frequently abandoning his family. Eliza compensated for any loneliness she may have felt with intellectual curiosity, and she became fascinated with the development of frontier democracy. This led to her first book, *Life in Prairie Land*, published in 1840. She also became involved in public debates about political issues, including women's rights and penal reform.

In 1844, Eliza Farnham was selected for the position of matron of the women's division of New York's Sing Sing prison. There, she introduced unprecedented reforms into a system that favored prisoner suffering over sympathy. She abolished the "no talking" rules, expanded the prison library, and provided literacy instruction. She offered opportunities for work and enhanced educational outreach through lectures and concerts. She ended barbaric customs such as punishing prisoners by wrapping them in blankets, handcuffing them, chaining them to the floor, and leaving them in public view. These reforms met with immediate condemnation by clergy and conservative critics, including the prison chaplain, for being too lenient, but others championed her efforts. Prominent novelist Catharine Sedgwick wrote that Eliza Farnham was, "of all women ever created (within my knowledge of God's works), the fittest for the enterprise. She has the nerves to explore alone the seven circles of Dante's Hell. She has physical strength and endurance, sound sense and philanthropy. . . . [She is] so rare a specimen of womanhood . . . this singular woman."

For four years, Eliza promoted improvements at Sing Sing, but, with the election of a new antireform prison board, she resigned and moved to Boston. She found work at the Perkins Institution caring for and educating deaf-mute students.

And then her husband died, and Eliza's life would never be the same.

A packet ship named the *Angelique* was engaged for the voyage of the California Association of American Women. Farnham's journey received much notice in the press:

> A lady of this State, well known for her labors in many a philanthropic cause, is about forming a benevolent expedition to California, which cannot but prove of great public benefit, in the present unsettled condition of that region. . . . She is now engaged in raising a company of intelligent and respectable females, to accompany her in this mission of charity. . . . This plan, the great utility of which will be seen at a glance, may be ranked among the most truly Christian enterprises of the day. Of the thousands of inexperienced emigrants now on their way to the land of promise, large numbers must feel severely the hardships and privations of their new mode of life, and their situation, without the aid thus furnished, must be indeed deplorable. We trust the originator of this expedition may meet with abundant support.
>
> *New York Tribune*, February 14, 1849

> Mrs. E. W. Farnham has just returned to the city, after visiting the Eastern States for the purpose of making up her company of migrating ladies, who, having no husbands to engage their attention here, are desirous of going on an errand to the golden land. The mission is a good one, and the projector deserves success. The enterprise in which Mrs. F. has engaged is one which evinces moral courage. Her reward will be found in the blessings which her countrymen will invoke for her when the vessel in which the association is to sail shall have arrived in California with her precious cargo. May favoring gales attend the good ship *Angelique*.
>
> *New York Herald*, April 12, 1849

New York was not alone in hailing the undertaking. Upon hearing the news, the *San Francisco Alta California* exclaimed: "For our own part, we are inclined to look favorably on the undertaking, as we have full confidence in the ability and integrity of Mrs. F. to conduct successfully such an enterprise."

But Eliza had difficulty getting subscribers for the undertaking, and as weeks passed she significantly lowered her expectations. In a letter to her friend Lydia Sigourney in April 1849, Eliza sadly testified: "My proposed expedition must be much smaller than I first desired to make it. It seems *probable* that it may now consist of 30 to 50; it is *possible* that it may not number 20." In the end the passenger list numbered twenty-two, with only a handful committed to the "California Association," including Eliza Farnham and her two sons.

On May 19, 1849, the *Angelique* cleared New York Harbor and headed for San Francisco. The ship had two major problems: dreadful drinking water and Phineas Windsor, a heartless captain who thoroughly disliked Eliza. At a stop in Valparaiso, Chile, while Eliza ran an onshore errand, the captain set sail, leaving her behind. Her children, however, were still on board the *Angelique*. As one witness wrote, "A brisk breeze was carrying us out to sea, leaving behind the mother; and it was truly a sight to try stout nerves to hear the children screaming and crying."

Meanwhile, news of the scaled-down expedition reached California. Sacramento's *Placer Times* reported on July 28, 1849: "We regret to hear that Mrs. Farnham's enterprise has turned out to be a sad failure."

In December the *Angelique* reached San Francisco, without Eliza. The *Daily Alta California* related that "the report . . . that the Angelique had arrived . . . produced a very decided sensation, as it was known that Mrs. F. had taken passage upon that vessel. We [wrote] a short paragraph [informing] our citizens upon the lady's arrival, but subsequently learned that the report was erroneous." Others who had waited hopefully for the *Angelique*'s arrival were likewise disappointed. William Redmond Ryan, a San Francisco resident, wrote: "The excitement was immense, and the disappointment proportionate, when the real facts became known. I verily believe there was more drunkenness, more gambling, more fighting, and more of everything that was bad, that night, than had ever before occurred in San Francisco within any similar space of time."

The *Alta California*, upon learning of the expedition's underwhelming response, and discounting Captain Windsor's callousness, dismissed the endeavor with a feeble attempt at humor:

> Her laudable endeavors to induce a large number of respectable young women to visit this country proved . . . futile. The will is always taken for the deed, and bachelors will unquestionably cherish the liveliest feelings of regard for the lady who so warmly exerted herself to bring a few spare-ribs to this market.

After securing passage from Valparaiso to San Francisco on another vessel, Eliza Farnham sued Captain Windsor for "breach of contract to convey plaintiff, her two children and servant from New York to San Francisco, and for ungentlemanly and unkind treatment of plaintiff and her children." Witnesses were called for the one-day trial, and Judge William B. Almond sent the case to the all-male jury without remarks or recommendation. When the court reconvened the next morning, Windsor was declared not guilty. The captain set sail on the *Angelique* that afternoon and left San Francisco.

Following the *Angelique* episode, Eliza Farnham settled her husband's estate, managed his Santa Cruz farm, and gathered her impressions for a book entitled *California, In-doors and Out*, published in 1856. During this time in California, she taught school, visited San Quentin Prison (which she found deficient), and lectured on such diverse topics as spiritualism, phrenology, and female valor.

Eliza married for a second and final time in March 1852 to William Fitzpatrick, who turned out to be a vicious drunkard. He beat her and threatened to kill her. A child, Mary, was born to the couple in 1853, but Mary died in 1855 of acute hydrocephalus. Eliza's ailing son, Eddie, had died a few months earlier. Eliza and William Fitzpatrick separated and divorced in June 1856.

Eliza then traveled to New York City, where she spent much of her time researching and writing her 1864 book *Woman and Her Era*, which focused on the social superiority of women. She had spoken on this topic before, at the National Women's Rights Convention in New York City in 1858.

Returning to California in 1859, Eliza Farnham lectured widely, and in 1861 and 1862 she was employed as the matron of the women's department of the Stockton insane asylum. With the onset of the Civil War, Eliza resettled on the East Coast, where she became active in the Women's Loyal National League advocating a constitutional amendment banning slavery. In July 1863, at the bloody pivot point of the Civil War, Eliza Farnham volunteered as a nurse at Gettysburg. While in Pennsylvania, she became ill and returned to New York City. Eliza Farnham died of "consumption," most likely tuberculosis, in 1864. She was forty-nine years old.

In 1893, Ora Kirby, the daughter of Eliza Farnham's close friend Georgiana Bruce Kirby, wrote this tribute:

> [Eliza Farnham] had a noble, courageous soul, full of sympathy for all misery and misfortune. Her own life had been a hard one. An orphaned childhood, desperate struggles for an education, with a mind which far outran the plodding conservative gait set for women of her day, to whom original thought meant original sin, she had to make her way in the face of constant condemnation from the stupid and uneducated frontier communities where fate had tossed her fortunes. . . . She bore herself with a pluck and pride which won respect even from those who did not respect her philosophies.

DRAPED IN A BUTTERFLY ROBE

THE MARIPOSA COUNTY COURTHOUSE

The California Gold Rush is continually studied, dissected, criticized, and celebrated, but direct evidence of this epic event is fading. There are ruins marked by roadside markers, lovingly maintained state historic parks, and the occasional period building repurposed for commercial use, but the physical manifestations of the era dwell mostly in memory. However, there are exceptions. Sprinkled here and there are Gold Rush–era structures still being used today as they were originally. The crown jewel is the magnificent Mariposa County Courthouse. It is a characterful structure that has sheltered many characters as well.

Constructed in 1854, Mariposa's is the oldest county courthouse in continuous use west of the Rockies. It may also be the most beautiful historic building in Gold Country. On its one hundredth anniversary, the California State Bar designated the courthouse a perpetual "shrine to justice in California."

When the courthouse was built, Mariposa County was much larger than it is today. Present-day California has fifty-eight counties, while Gold Rush California had only twenty-seven—and Mariposa was the largest. In the 1850s, the county occupied more than one-fifth of the state's acreage, or thirty thousand square miles. To meet the needs of the exploding Gold Rush population, construction of a courthouse was authorized by the Mariposa County Court of Sessions, and the firm of Perrin Fox and Augustus Shriver was given $9,000 and one year to build the New England–style two-story structure. Fox and Shriver used a nearby stand of white pine as their material and they set to work. The lumber was cut on site and hand-planed. Nails were not used in the supporting structure, as the courthouse is fitted together with tongue-and-groove lumber, mortise and tenon joints, and pine pegs.

Still in use today, the second-story courtroom rests on plank floors, and the jury box, spectators' benches, judge's bench, and counsel tables all show unmistakable 160-year-old woodworking marks. A potbellied stove warms the litigants. It is a serviceable courtroom, but it is also a museum to the practice of law in the early Gold Rush days. Footsteps reverberate in the old building, and the echoes of a bygone legal era are just as strong. Modern amenities, such as electrical lights and climate control systems, are hidden from view, and a narrow staircase, originally constructed so that a single armed guard could easily defend the entrance to the first floor, drops to a museum that features artifacts, framed historical documents, portraits of solemn former judges, and a collection of cattle brands. In 1866, a tower was added and a clock for it was shipped around Cape Horn at a cost of $1,130.35. The clock was most likely constructed in England and is wound by cranking two weighted cables onto drums. Some visitors swear that the courthouse shakes when the clock chimes.

The enchanting structure brought architectural grace, dignity, and decorum to what were often haphazard and remarkably rapid trials in the county. In 1890, prominent historian Hubert Howe Bancroft recalled a Mariposa trial of 1850, years before the courthouse was built. "Court was held under a tree," Bancroft wrote, "and the jury retired to another tree to deliberate." The case concerned the assault and wounding of one miner by another. In the shade, Bancroft recounted, the jury deliberations commenced:

> "Let's hang him," said [juror] number one.
>
> "Oh, no," replied number two, "he only stabbed a man; we can't hang him for that."
>
> "Send him to the state prison for life," put in number three.

"That'll do," exclaimed half a dozen at once. And so it was concluded, all agreeing to it.

"It seems rather hard after all," ruminated number two, as the twelve started back for the court-tree, "to imprison a man for life, for merely stabbing a [man]; besides, where is your prison?"

"Let's acquit him," said number one.

"Agreed," exclaimed the rest; and so the man was set at liberty.

The Mariposa Courthouse has been the venue for some of the most notable civil, water, and mining cases in California history. Arguably the most important and celebrity-laden was the 1857 case of *Biddle Boggs v. Merced Mining Company.*

Mariposa County takes its name from the massive Mexican land grant Rancho Las Mariposas, in the Sierra Nevada foothills. The 44,387-acre parcel was bestowed to Juan Bautista Alvarado in 1844 by Governor Manuel Micheltorena and then sold to John C. Frémont, "the Great Pathfinder of the West," in 1847. But when California was transferred to United States authority, those who possessed Mexican land grants had to prove their ownership in court. In 1852, based upon his own private survey, Frémont was confirmed as the owner of Rancho Las Mariposas, although some disagreed and the decision was appealed and reversed. Frémont took his case to the United States Supreme Court, and in December 1854 the court declared Frémont's claim valid and ordered an official survey. In 1856, with all the legalities completed, John C. Frémont's Rancho Las Mariposas was patented in his name.

While that legal battle unfurled, Frémont's rich land was overrun by gold seekers. Once his ownership was affirmed, he leased a portion of his land to a miner named Biddle Boggs, but as it turned out, that particular real estate was already occupied by the Merced Mining Company. The company had been working the property since 1851, under permission from the county assayer, and they argued they were on government land, not Frémont land. The Merced Mining Company had invested more than $800,000 in mine buildings, mills, and

The Mariposa County Courthouse, shown here in 1936, has changed little since its construction in 1854. Courtesy of the Library of Congress, Prints and Photographs Division, Washington D.C.; Historic American Buildings Survey photo by Wm. H. Knowles, May 1936; HABS CA-1105, ID: HABS CAL,22-MARI-1.

machinery there. They sued over the lease to Boggs and asserted that Frémont had no claim to any mineral rights. Starting in May 1857 and lasting for weeks, the trial took place in the beautiful courthouse in Mariposa. This proceeding would have long-term consequences.

Featuring not only the famous Frémont, who had recently been the first presidential candidate of the new Republican Party, three of the lawyers on the case later went on to serve on the California Supreme Court. In July 1857, District Judge Edward Burke of the Thirteenth Judicial District ruled in favor of Frémont, but in 1858, the California Supreme Court reversed the decision and ruled in favor of the Merced Mining Company. Leading this latter effort was Chief Justice David Terry. In an unusual move, a rehearing was granted, and in 1859 the California Supreme Court reversed itself and found in favor of Boggs and Frémont. The Merced Mining Company lost its investment and capital improvements. Frémont had spent seven years and hundreds of thousands of dollars in litigation, but his property rights had been affirmed once again.

Issuing the California Supreme Court's decision was newly appointed chief justice Stephen Field, a forty-niner from Connecticut. Three days after arriving in California in December 1849, Field was elected alcalde of Marysville, where he was best known during his brief tenure for setting up a whipping post for summary punishment for lawbreakers. After serving one term in the state legislature, he returned to private law practice and, in 1857, was elevated to the California Supreme Court, becoming a chief justice in 1858 following David Terry's reelection defeat. As a justice, Field was so thoroughly disliked that, as a precaution in case of attack, he had a special coat tailored with two hidden pockets for pistols. In 1858, he was challenged to a duel by fellow justice William Barbour, which thankfully ended bloodlessly when neither man fired his weapon.

In his decision in *Biddle Boggs v. Merced Mining Company* Field sided with Frémont. The chief justice publicly noted that the Great Pathfinder probably did not deserve to win but that the law compelled the court to find in his favor. On November 26, 1859, a correspondent to the *Daily Alta California* sarcastically summarized Field's decision regarding the mine property and mineral rights:

> Yes, take them; they are yours. You did not claim the land when these people were building the works, in fact you disclaimed it, but that was a mistake for which you must not be prejudiced; you did not find the vein or make the tunnel, or build the mills, or level the roads; you have done nothing

> except manage a survey skillfully; but take the property, the law gives it to you.

John C. Frémont emerged triumphant in this long court case, but the victory was fleeting. His legal battles and poor financial decisions left him deeply in debt. Represented by Chief Justice Field's brother David, Frémont sold his Rancho Las Mariposas to New York investors in 1863 to satisfy his creditors, and he never set foot in Mariposa again. Frémont's detractors exulted in his downfall. Critics had claimed that Justice Field had been bribed by Frémont to secure a favorable outcome (Field and his brother were accused of corruption in a nasty pamphlet entitled "The Gold Key Court, or the Corruptions of a Majority of It"), but no evidence of a conspiracy was ever found.

By 1863, Stephen Field had moved on. In that year, Abraham Lincoln selected him for the Supreme Court of the United States, where he served until his resignation in 1897. During his tenure, Field continued to antagonize his rivals, however, and in 1889 it caught up with him when he was the target of an alleged assassination attempt by David Terry, who had preceded Field as California Supreme Court chief justice until he fled the state after killing David Broderick, the U.S. senator from California. During an unexpected encounter at a railroad station, Terry supposedly assaulted Field and was then shot and killed by Field's bodyguard. Many believe Terry's loathing of Field began during the *Biddle Boggs v. Merced Mining Company* proceedings in the picturesque courthouse in Mariposa.

In January 1968, the *American Bar Association Journal* profiled the Mariposa County Courthouse. It concluded:

> Like all frontier mining communities, Mariposa had its share of violence tempered by sentiment, claim-jumping accompanied by gun-fighting, justice produced by expediency, and of the confusion and racial brawling that usually result when men of many nations and cultures converge. . . . [The courthouse's] calm New England exterior is a contrast to the era of turbulence that produced it.

"CONVULSIVE THROES AND CONFLICTS OF PASSION"

DR. JOHN MORSE, THE FIRST HISTORIAN

When the Gold Rush began, the interior of California was largely unmapped and unfamiliar, but many concluded, sight unseen, that the environment must be identical to that of the pleasant Pacific coastline. A few even believed the Sierra Nevada region was tropical. Frequently, the uninformed touted the curative benefits of the Mother Lode, as did, for example, Jacques Antoine Moerenhout, the French consul in Monterey, who on July 30, 1848, reported to his superiors that the foothills, which he had never visited, possessed a "mild healthful climate." Pringle Shaw, an early immigrant from Canada, gushed that "the climate of the main portion of California, may be reckoned, . . . as among the most delightful on the globe. . . . Consumptions, rheumatism, fevers, and every sort of epidemic, are almost unknown." Others echoed these sentiments, and based on such apparently authoritative testimonies, it became commonplace for East Coast doctors to recommend a period spent in the California goldfields as a remedy for acute ailments. So when Dr. John Morse fell ill in New York in early 1849, his doctor prescribed a restorative residence in the golden land across the continent.

John Frederick Morse was born in Essex, Vermont, on December 27, 1815. In 1843, he married Rebecca Canmore of Connecticut in New York City, where he had attended medical school at the University of the City of New York (now NYU). After graduation in 1842, Morse opened a successful medical practice in Brooklyn. His daughter Emma was born in 1846. Life was fulfilling and his prospects were excellent. Then he became ill and his doctor issued the California prescription, at which point Morse booked passage for Panama on the bark *Begota* and left New York. Crossing the Isthmus of Panama, Morse secured passage to San Francisco on the *Alexander von Humboldt*, arriving in August 1849. Dr. Morse initially settled in Sacramento and then headed to the diggings in Coloma, where he failed as a miner. He returned to Sacramento and resumed his medical career.

Morse immediately discovered that there was plenty for him to do in Gold Rush Sacramento. With people pouring in from every direction every day, the city was an incubator for disease, and in addition to several notable epidemics, Sacramento was also repeatedly afflicted with floods and fires in its early years. With Dr. J. D. B. Stillman, Morse opened an office and hospital near the Embarcadero, and, in association with others, was instrumental in establishing other hospitals in the boomtown. In May 1850, John Morse established the Medico-Chirurgical Association, the first medical society in California. This society initiated state licensing of physicians, codified sanitation standards, and inaugurated the first system of health insurance in California by funding the care of indigent patients.

To make ends meet in the rapidly changing economy of Sacramento, Morse also busied himself as a real estate agent, an auctioneer, and the editor of the *Sacramento Union*. Rebecca and Emma joined him in 1851, and with his career progressing and his family once again whole, his path was secure and promising. And then, as it so often did, disaster struck.

On the morning of November 3, 1852, thirty-six-year-old Dr. John Frederick Morse was emotionally paralyzed. The night before, the Great Sacramento Fire had destroyed nearly 90 percent of the city and leveled more than twenty-five hundred structures, including his medical office. But the fire was not what troubled Morse that morning. His wife, Rebecca, nine months pregnant, and his daughter, Emma, only six years old, were terrified. John hustled them both to the still

Sacramento physician Dr. John Morse wrote what is considered to be the first history of the Gold Rush. Courtesy of the California State Library, Sacramento; California History Section.

smoldering waterfront on the Sacramento River and escorted them onto the steamboat *Comanche,* for passage to the safety of San Francisco. He bid them farewell, returned to the dock, and watched as the *Comanche* drifted away. It was the last time he saw his wife alive.

The next day, November 4, Rebecca went into labor, and aboard the steamboat she gave birth to a son and died shortly after. Several days passed before Dr. Morse learned he had a son and that his wife had tragically died in childbirth. He was devastated. He had lost his wife, his business had been destroyed, and he was separated from his daughter and newborn son at a moment of calamity. He eventually reunited with his young children, but the heartache was profound, and in what could fairly be characterized as therapy, Dr. Morse plunged into work and service. As he mended, John Frederick Morse wrote the first published history of Sacramento. It is widely considered the inaugural attempt to provide not merely a personal recollection but a historical perspective on the California Gold Rush. Dr. Morse could be justly designated the first historian of the Age of Gold.

He did not come to the project independently. In 1853, Samuel Colville, publisher of the *Sacramento Directory*, a listing of city residents and businesses, asked Morse to write a history of Sacramento from 1849 to 1853 for inclusion in the 1853–54 directory. Though Morse had journalistic experience and was a prominent and active member of the community, he had no training as a historian, and it showed; his forty-page history is not an exemplar of modern historical scholarship. There is no pretense of objectivity, key events are omitted, and Morse was prone to bursts of impassioned judgment throughout. He acknowledged the work's deficiencies in the note he wrote to Samuel Colville when he delivered the manuscript in October 1853: "I should be most unwilling to have it presented as a well digested and carefully written history."

But for all of its shortcomings, Morse's history is nevertheless an insightful and evocative account by a man with a singular perspective. Morse was, after all, a doctor, not a historian, and his concern for the physical and moral health of Sacramento imbues his work. In so doing it swings wide the doors of a bygone era so that we can steal a glimpse of a tumultuous—and frequently deadly—time and place.

Dr. John Morse opened his chronicle of the river city in early 1849, a time when hopes were high but the population was low.

> The old pioneers and newly arrived adventurers constituted, at that time, but a small and insignificant community, and whilst they were fully impressed with an idea of the profusion of riches that surrounded them, they had not as yet a conception of the convulsive throes and conflicts of passion to which a pursuit of California gold must inevitably lead.

Morse was impressed with the "marvelous spirit of honesty" these early gold seekers displayed where "neither goods nor gold dust were watched with the least care of consideration." He gazed in astonishment at "miners coming to town, freighted with bags of the valuable ore, [who] stowed away their treasure as indifferently as they did their hats and boots."

By May 1849, argonauts began arriving in force, and soon "thousands and thousands" were using Sacramento as a staging point for their adventures in Gold Country. It was far from an organized activity, however:

> Sacramento then began her career as a most important trading center. It was the grand starting point to every new coming gold seeker . . . and made Sacramento the peculiar town for the purchase of gold digging implements and provisions. In June, every thing in and about the city indicated an overwhelming business, conducted without a particle of method, and in such utter confusion and recklessness of manner as to make it impossible for a man to construct calculations that embraced more than the contingencies of a single day.

Morse was among the first to note the pattern that gambling halls were often the earliest establishments constructed in mining camps. Perhaps snidely, he commented that, in Sacramento's case, "public gambling became one of the leading and absorbing features of our city's progress and *greatness*." Morse maintained that while merchants, bankers, and more conservative figures bided their time and money before committing to building, "gamesters were viewing [vying] with each other in the erection of magnificent saloons, at an expense that would startle credulity and pervade the soul of a reflecting man with a shudder." Sacramento's first gambling saloon was far from magnificent: "A few poles stuck in the ground, and covered with a windsail, constituted the first gaming rendezvous, and bore the very appropriate name of 'Stinking Tent.'"

Dr. Morse unleashed his vitriol on those who frequented the gaming parlors, which he characterized as "the polluting altar of unproductive and self destroying avocations" and infernal dens of "naked, unmasked depravity."

> Hundreds and thousands of men, who had been reared to regard gambling as a stain upon the character of a man, who had left their homes by means of borrowed money, and left behind them women and children to toil for their subsistence, until the golden dreams of California should be realized—hundreds and thousands of such men could be seen crowding these miserable haunts of ruin, and gambling away the first hundred or thousand dollars which they had made in the country.

Other commercial interests caught his eye as well. Morse noticed the enormous profits gained from "mining the miners." He listed the high prices of commodities, fees for services that were ten times or more the going rate on the East Coast, and the roaring free enterprise, with its *caveat emptor* nature, of the Gold Rush economy. He remarked that "teaming and packing goods to the mines were employments which absorbed a tremendous amount of energy and enterprise, and yielded princely revenues." And the beehive of this buzz of activity was flourishing Sacramento City, the "nucleus of attraction to the world":

> It was the great starting point to the vast and glittering gold fields of California, with the tales of which the whole universe became astounded and which men of every clime and nation sought to reach without a moment's reflection upon the cost or hazard of such an adventure. The only consideration

> upon the part of a hundred thousand gold seekers who were preparing for immigration to California was dispatch.

As a physician, John Morse was more aware of the human cost of poor sanitation, indifferent hygiene, disease, and natural catastrophe than most pioneer chroniclers. He described a blanket-wrapped corpse delivered to his hospital during the massive flood of 1850:

> Fortunately for him, death was the speedy alternative. His troubles were ended. A finely developed form, a face upon which lingered the indices of cultivated intellect, a heart that once beat with manly pride, were unwrapped in a death so dreadful as to beggar description, and so appalling as to excite an almost eternal impression of nausea and disgust in the minds of those who beheld it. The blanket was with difficulty detached and when drawn off presented a shirtless body already partially devoured by an immense bed of maggots occupying nearly as much space as the emaciated carcass itself. And when one adds to this loathsome mass, these crawling elements of disgust, the accumulated excretions which were alike confined by the agglutinated folds of the blanket, a head of hair almost clogged up with vermin, then can a just conception be formed of what was suffered during the sickness of the fall and winter of '49.

The same flood evoked this report on the lighthearted attitude—perhaps a form of denial—of many in the immediate neighborhood of the gruesome mortality and devastation:

> At 10 o'clock on the evening of the flood, when the back waters of the slough and the waters that came pouring in from the banks of the Sacramento, were rushing into the city, tearing up side-walks and dislodging merchandise, sweeping away tents and upsetting houses; at this very time, and throughout the inundation the city seemed almost mad with boisterous frolic, with the most irresistible disposition to reel in all the joking, laughing, talking, drinking, swearing, dancing and shouting.

Death lurks in the shadows even when Morse describes joyous times. Here he links California's admission as a state in 1850 to a cholera epidemic that killed 10 percent of Sacramento's population.

> Early in the morning of the 15th of October . . . the booming notes of a rapidly fired cannon upon the Levee waked the citizens of Sacramento with the understood and emphatic assurances of ADMISSION. . . . But, alas, the exuberance of spirit thus enkindled, the joyous and buoyant feelings thus excited were but the illusive precedents of one of the most appalling calamities that had ever yet set its seals of distress upon the destiny of the valley city. Associated with the glorious intelligence of our admission into the great confederation of States, was the sad assurance that a most malignant cholera was sweeping on towards California, and that the passengers on the very steamer that brought the news had many of them fallen to this terrific scourge.

The cholera epidemic spread throughout the state and claimed somewhere between six hundred and one thousand victims. In Sacramento, seventeen doctors died fighting the disease. Morse's partner, Dr. Stillman, called the outbreak the "season of death."

> This awful calamity lasted in its malignant form but about twenty days, but by the unsystematic records of the times the number of deaths cannot be ascertained. Besides those that died in the city, many were overtaken by death in other places, and upon the road, in the desperate efforts of our citizens to escape by running from the enemy. . . . By the time that the disease had almost completely disappeared, the City was nearly depopulated, and there were not a few outside croakers [observers], that intimated that the Levee City was dead beyond a possibility of resurrection.

Morse concluded with a tribute to the hardiness of Sacramento residents, among the first accolades recording the resilient spirit of the gold seekers manifested throughout Gold Country:

> But those who suppose Sacramento and Sacramentans could be so easily crushed, had not learned their character. The very moment that mortality began an obvious retreat from our premises, that very moment that those who survived their flight returned, and those who abided by the City in its distress, reacted upon the calamities of the town with such an elastic and vigorous energy as to completely transform the appearance of the place in a few days . . . [Sacramento's] future is no longer in doubt, but a certainty.

Dr. Morse remarried a few months after his history was published and, with his second wife, Caroline, he had four more children, a boy and three girls. Heartbreakingly, his son by Rebecca, John Francis, who had been born on the steamboat *Comanche* in November 1852, died at the age of four.

Over the following decade Morse continued to be a prominent and civic-minded Sacramento resident. He was instrumental in establishing libraries in Sacramento and, as the representative of the Society of California Pioneers, he spoke at the Sacramento groundbreaking ceremony for the transcontinental railroad on January 8, 1863.

In late 1863, Morse moved to San Francisco to teach at the first medical college in California. In 1874, his health failing, he boarded a ship to Australia in search of a cure. His doctors believed his condition would improve with a sea voyage, but, days into the journey, his ailment worsened and he returned to San Francisco. Dr. John Morse died on December 30, 1874, three days after his fifty-ninth birthday.

CHEESQUATALAWNY
JOHN ROLLIN RIDGE

In the quiet Greenwood Cemetery in Grass Valley, there is a row of markers for the Ridge family. Most prominent among them is the headstone for John Rollin Ridge, one of the most interesting and influential figures of Gold Rush California. He died at the age of forty in 1867, and the name he had when he died was not the one he had been given at birth.

John Rollin Ridge, the son of a powerful family in the Cherokee Nation, was born in 1827 near today's Rome, Georgia. His birth name, Cheesquatalawny, means "Yellow Bird." When he was just three years old, he endured one of the most traumatic moments in the tribe's history when the Indian Removal Act accelerated the eviction of Cherokees from their homes in the Southeast. The Cherokees resisted, but federal officials exploited disagreements within the tribal community over how to proceed, and in 1835 the government convinced twenty-one Cherokees, including John Rollin's grandfather, Major Ridge, and John Rollin's father, John Ridge, to sign the Treaty of New Echota. The treaty banished tribal members to the West, forcing them to abandon all Cherokee lands east of the Mississippi. Some Cherokees supported the treaty, but others felt it was a betrayal. A portion of the removal to Indian Territory (today's Oklahoma), during which thousands of Cherokees died, came to be called the Trail of Tears. The Cherokee name for the event translates as "The Trail Upon Which We Cried."

Tensions within the tribe intensified following the treaty, and within months of removal, personal antagonisms burst into violence. In 1839, Major Ridge and John Ridge were murdered by other Cherokees. Twelve-year-old John Rollin witnessed his father's assassination.

John Rollin Ridge left Indian Territory immediately and traveled to Arkansas, where he lived for four years. In 1843, young John Rollin went to Massachusetts for schooling, returning to Arkansas in 1845 and beginning his law practice. John Rollin was particularly interested in Cherokee politics and closely followed the developments within the tribe. On more than one occasion, he expressed a desire to avenge the deaths of his father and grandfather. In 1847, he married Elizabeth Wilson, a white woman he had met in Massachusetts, and one year later the couple had their only child, Alice Bird.

As John Rollin's involvement in tribal politics deepened, he grew increasingly angry and zealous, and, in 1849, his passion boiled over into bloodshed. John Rollin killed David Kell, a Cherokee man he believed was one of his father's assassins.

Largely to escape prosecution, John Rollin bolted and joined the rush to California. He left his young family behind, promising to reunite when circumstances permitted. Ridge, a slaveowner, sold one of his slaves to finance the journey. Accompanied by his younger brother Aeneas, Ridge joined the wagon company of Major Elias Rector and Colonel Matthew Leeper in April 1850 and followed the Oregon-California Trail to California. As did many who made the lengthy trip to the goldfields, Ridge encountered difficulties on the trail, some of which he detailed in letters back home. In a letter to his mother on October 4, 1850, he wrote that water and feed for his horses and mules was scarce as they approached Wyoming, and, concerned for the animals' welfare, he lightened the wagonload to the bare essentials, discarding all his possessions except for an extra pair of pants and a frying pan. He drank polluted water and suffered a bout of diarrhea, his money was extorted by crooked ferrymen and traders, and as his food stores spoiled in the sun, he found himself constantly "faint and hungry" and eating meals that featured "meat that tasted like rust, and a piece of bread that made the stomach retch at every swallow." As he told his mother, he ran out of money and sold a pony to purchase what he thought would be sufficient provisions to reach California. He was wrong.

Upon reaching the Forty Mile Desert in northern Nevada, he recounted, Ridge combatted alkali dust four inches deep. The caustic dust produced sores in his nose, ears, and throat and severely damaged the hooves of his pack animals. Once again, the brothers ran out of food and could not afford to buy more. John Rollin and Aeneas trekked across the burning desert without food and water. As he wrote to his mother, her sons could not keep clean and eventually just stopped trying. Ridge recalled that his hair resembled a matted, dirty, and stinking mop used to scrub floors. Surviving this harrowing gauntlet, they limped into Truckee Meadows, near today's Reno. John Rollin sold a mule for $11, which was just enough to buy that night's dinner.

Ridge was bemused at the extraordinary, perhaps foolhardy, efforts that he and Aeneas were enduring to reach the golden promise of California. "Poor Humanity!" he jokingly wrote his mother, "to what miserable passes will it put itself for money."

Finally, after months on the trail, Ridge and his brother arrived in the promised land of Gold Rush California. In the autumn of 1850, they entered Placerville, then called Hangtown. After selling his best mule for $45, Ridge purchased a few days' worth of food and hay for his remaining handful of pack animals, and then asked about mining prospects in the district. The answer was discouraging. He was told that "thousands were already digging in the town itself and at every little hole for six miles, up and down the creek. There was not a solitary place to dig. Every claim was taken." He separated from his brother Aeneas and moved on.

John Rollin Ridge mined in other regions of Gold Country, at one point even leaving the Mother Lode and attempting placer mining along the Trinity River in the far northwestern reaches of California and near Mt. Shasta. No luck. Eventually, the endless physical labor and pathetic results led Ridge to abandon mining altogether. Enough was enough. In his long October 1850 letter to his mother, he described his last day as a miner: after standing knee deep in Shasta County mud, pelted by nonstop rain and snow, and having realized only fifty cents in gold dust from hours of panning, John Rollin simply threw down his pick and pan and walked away. He sold his remaining stock animals for $75 and headed on foot to Sacramento, two hundred miles away.

John Rollin Ridge found himself in boomtown Sacramento nearly penniless and "willing to do anything honest," as he later recalled. His experiences in the diggings led to the revelation that the most fruitful path to success was by providing a service to eager and improvident miners. Ridge felt he possessed some ability as a writer, and on the recommendation of a friend, he approached the eccentric Colonel Joseph Grant, a well-known Sacramento street orator and California agent for the *New Orleans True Delta*, a popular journal. John Rollin asked Grant for a job as a correspondent for the publication and, after writing a test article for Grant, Ridge was hired at $8 per report.

John Rollin Ridge quickly gained a reputation as a first-class newspaper editor, reporter, and columnist. In 1852, he was reunited with Elizabeth Wilson Ridge and Alice Bird, who had made the long journey to the goldfields via the Isthmus of Panama. From 1852 to 1864, the Ridge family lived in six different California communities, including Sacramento and San Francisco, as he pursued his newspaper career. Ridge worked as an editor for several California newspapers, including the *California Express*, the *National Democrat*, the *San Francisco Herald*, and the *Red Bluff Beacon*, and he was one of the founders of the *Sacramento Bee*. He wrote poetry on the side, including one of the earliest published works about Mt. Shasta.

Not surprisingly, John Rollin wrote extensively about Native American politics. But surprisingly, he was often scathing toward the Indian communities themselves. He disagreed with the notion that Indians should remain independent from government control, believing that the federal government provided necessary guidance and assistance to the tribes. He felt California Indians were inferior to other Natives, and he tacitly supported policies that stripped California Natives of their lands and rights, although he decried the violent efforts to exterminate the Indian population. As a former slaveowner, John Rollin was also sympathetic to the conservative faction of the Democratic Party that advocated slavery extension to California, and with the advent of the Civil War, his writings were a study in contradictions. He supported retaining national unity at all costs, but he also protested the election of Abraham Lincoln and was supportive of some aspects of the Confederacy.

For all his contributions to journalism in California in the decades following the Gold Rush, John Rollin Ridge left his greatest mark with a work of fiction. In 1854, he wrote a book about a sensational California bandit based upon the exploits of several real-life outlaws who took the name Joaquín, including one named Joaquín Murieta. Ridge's fertile imagination consolidated the stories into a single myth under that one man's name, and he exaggerated details, invented heart-pounding situations, concocted wild escapes, and promoted the image of Joaquín as a Mother Lode Robin Hood driven to crime by social injustice. Ridge's tale also offers parallels to his own life story of cultural duplicity, a quest for vengeance, and the complexities of navigating between two cultures. As James Parins, John Rollin Ridge's biographer, wrote: "If Ridge can be said to have created Joaquín, he did so in his own image." Even Ridge's physical description of Murieta closely matched the author's likeness. Less than one hundred pages, *The Life and Adventures of Joaquín Murieta, the Celebrated California Bandit* is generally considered to be both the first novel written by a Native American and the first novel published in California.

In 1864, John Rollin Ridge and family, including another younger brother, Andrew Jackson Ridge, moved to Grass Valley, Nevada County. John Rollin purchased an interest in the *Grass Valley National* newspaper, which he coedited with W. S. Bryne.

In 1866, Ridge briefly traveled to Washington, D.C., as part of a Cherokee delegation hoping to annex Indian Territory into the union as a state. The effort failed and Ridge returned to the downtown Grass Valley home he shared with his family. He fell ill from "brain fever," most likely encephalitis, and died on October 5, 1867. In his October 8 obituary, it was written:

> As a writer probably no man in California had a wider and better reputation than John R. Ridge. He possessed a good

> education, had a clear and vigorous mind, was well up in classical lore; and in the possession of these essentials to journalistic distinction it is not surprising that he was professionally successful. With more energy and with stronger aspirations to place his name among the highest literary lights he might have added many volumes to the purer and better literature of the time. . . . He wrote with ease, and as is generally the case with genius, sometimes carelessly. . . . His remains were yesterday interred in Greenwood Cemetery near this place, his funeral cortege being a very large one.

John Rollin Ridge rests in final slumber next to his wife, Elizabeth; his daughter, Alice; his brother, Andrew Jackson Ridge; and some in-laws.

In 1876, his widow planted a red maple tree at the corner of School and Neal Street in honor of her husband. The tree came from the battlefield at Gettysburg. Although it has been seriously damaged by powerful storms over the decades, it still stands today.

John Rollin Ridge lives on through his stories of Joaquín Murieta, the legendary bandit hero of California Gold Country. The mythology surrounding this colorful outlaw stubbornly refuses to expire, and his name is still invoked throughout the Mother Lode. Sprinkled throughout the region are plaques, inscriptions, and markers recounting the prodigious feats of Murieta, "our" Joaquín.

Many scholars believe that Ridge's mix of fact and fancy influenced later writers, such as New York's pulp writer Johnston McCulley. In 1919, McCulley grafted Ridge's Joaquín Murieta legends onto additional factual accounts of Californio outlaws Salomon Maria Simeon Pico and Domingo Hernandez and crafted "The Curse of Capistrano," a magazine serial introducing the character Zorro. Zorro shared many characteristics with Murieta, including a swashbuckling bravura and a palpable sense of social justice, and other commentators have noticed similarities between Zorro and Batman, a masked avenger who lives a dual life of publicly opulent gentility on the one hand and stealthy, rugged action on the other. Not many modern Americans would guess that this heroic literary continuum, now stapled firmly into our

John Rollin Ridge, a Native American author and journalist, wrote the first novel published in California: the story of legendary Gold Rush outlaw Joaquín Murieta. Courtesy of the California State Library, Sacramento; California History Section.

popular entertainment and cultural lineage, can trace its origins to the intriguingly multifaceted figure of Cheesquatalawny and his ninety-one-page Gold Rush adventure story.

John Rollin Ridge was an enigma, a sojourner in dual universes. As James Parins concludes in his excellent 1991 biography *John Rollin Ridge: His Life and Works*: "[Ridge] found himself caught between two worlds, discovering sympathy and hostility in both. . . . John Rollin Ridge knew an inner conflict as well, which found his Indian identity at odds with the need to survive in a white-dominated society. He never fully resolved that conflict."

"I AM JOAQUÍN!"

THE LEGACY OF JOAQUÍN MURIETA, THE CELEBRATED CALIFORNIA BANDIT

He was crafty, generous, vindictive, heroic, and remarkably cool under pressure. For some, he was a kindly benefactor, champion of the oppressed. For others, he was a cold-blooded killer. He was a man of startling handsomeness and bravado. He was a stealthy avenger, a resourceful escape artist, a loyal friend of the downtrodden, and the swashbuckling defender of a lost culture. He was Joaquín Murieta, the bandit hero of Gold Country. And he was everywhere. In Sawmill Flat, outside the Southern Mother Lode town of Sonora, they boast of being the location of Joaquín's first homestead after he had arrived from Sonora, Mexico, in 1850. Down the road in Murphys, they tell the tale of the brave young Joaquín swearing revenge on the Anglo hooligans who tied him to a tree, beat him bloody, raped his wife, Rosa, and killed his half-brother. A few miles away, in San Andreas, they tell the story of Joaquín's bulletproof vest, constructed for him by a sympathetic French argonaut: to test its effectiveness (for Joaquín was grateful and yet practical), he made the Frenchman wear the vest while Joaquín shot at him. In Hornitos, they point to a tunnel that Joaquín supposedly used to escape from heavily armed pursuers, and in Volcano, they claim that Joaquín once sought refuge in a well-hidden tree house as tired lawmen rested below him. Near Stockton, the story is told of how Joaquín's band commandeered a ferryboat, but upon learning that the captain of the vessel had only a few dollars in his pocket, Joaquín paid the poor man handsomely for the transport and his trouble. Dozens of towns claim that Joaquín camped here, Joaquín escaped here, Joaquín slept here, Joaquín assisted the needy here, Joaquín was here!

But was he real?

His supporters steadfastly proclaimed he was, while his detractors sniffed that he was nothing but the creation of a dime novelist's fevered imagination. There are a handful of believers who claim that every incident, every nuance is true, all true. Some admit that the facts may have been fudged a mite, but the skeleton of truth is secure. A few historians acknowledge that some proof can be found in the historical record, but, for the most part, the story of Joaquín Murieta (sometime spelled Murrieta or Murietta) is an exaggeration. Many scholars believe that, all things considered, the Murieta tale for public consumption is myth, an entertaining yarn constructed of whole cloth, smoke, and mirrors.

Those who argue that the Murieta case is a compelling mixture of fact and fable rely on historian Frank Latta for evidence. Latta, author of the 1980 *Joaquín Murrieta and His Horse Gangs*, sought the "historical Joaquín" for decades. He uncovered persuasive accounts that largely conform to the many legends of Joaquín albeit minus the fantastic, superheroic elements. Latta spells out the story of a proud Murieta family that had been in Mexico and the Alta California province for generations. In 1849, when the Gold Rush was in ascendancy and the systematic divestiture of Californios commenced, young Joaquín, then in his early twenties, was driven from a rich mining claim in Niles Canyon, then part of Contra Costa County. His wife, Rosa, was raped and his half-brother Jesús was lynched, and the grief-stricken Joaquín was severely beaten. In spite of these events, he gravitated toward the teeming mining camps and may have worked in a gambling hall as a montebank dealer. He later toiled as a horse dealer and then probably became a horse thief and, possibly, an outlaw leader of a gang.

We also know that by 1852 or 1853 the Southern Mines were troubled by a number of bothersome, elusive, and sometimes murderous outlaws. In the wake of the 1850 Foreign Miners' Tax, which required non-English-speaking immigrants to pay a monthly mining fee, and the expulsion of Californios and Mexicans from the goldfields, land pirates began preying on the mining communities. While many of the outlaws were known to be of European heritage, the majority were Spanish-speaking, presumably from Mexico. Crime increased dramatically and fear mounted across Gold Country.

Most of these bandits were named—or claimed to be named—Joaquín, although the last names varied. There was a Murieta, but also a Valenzuela, Carrillo, Botellier, and Ocomorenia. The perpetrator was often identified only as "Joaquín" by those interviewed at the scene, so it was difficult, if not downright impossible, to determine which Joaquín had committed which crime.

But despite not knowing who was responsible for the crime spree, or even how many people were involved, the mining communities were desperate for it to come to an end. In May 1853, the California legislature passed "An Act to Authorize the Raising of a Company of Rangers" and hired gunman and former Texas Ranger Harry Love to capture Joaquín dead or alive. Five last names were listed as possibilities, including the creatively spelled "Muriata." As an inducement, Governor John Bigler offered Love and his twenty-member posse a $1,000 reward.

Love and his compatriots rode out to seize Joaquín and his reported accomplice, Three-Fingered Jack, the alias of Manuel Garcia, who was wanted throughout California for theft and murder and was a suspected serial killer of Chinese miners in Calaveras County. It was widely rumored that Joaquín Murieta had killed as many as two hundred Chinese people as well. The pressure on Love and his crew to deliver Joaquín was intense.

For several weeks, they had no luck. Then, near Panoche Pass, about fifty miles from Monterey, Love's patrol encountered a band of Mexicans. They killed at least two of them, one of whom had claimed to be the group's leader but had never mentioned his name. Love decapitated this supposed chieftain and had his head sealed in a jar of alcohol; in a separate jar he kept the severed hand of a second victim. Love claimed the head was that of Joaquín Murieta and the hand had once belonged to Three-Fingered Jack. He returned triumphantly to Fresno to claim his reward.

But many were skeptical, to say the least. A surviving member of the Mexican party stated that the head was clearly that of Joaquín Valenzuela, not Joaquín Murieta, and witnesses to his various crimes who saw the gruesome pickled head swore that it bore no resemblance to the "actual" Joaquín. Love insisted that the grotesque relics were genuine and exhibited seventeen sworn affidavits, including one from a Catholic priest, attesting to their authenticity. He exhibited his grisly trophies throughout Gold Country for a $1 fee, although never in Calaveras County, where the "real" Joaquín Murieta reportedly kept his primary hangout.

Initially Love's exhibition drew large crowds, but by 1856, interest was dwindling. In that year, a San Francisco entrepreneur purchased the head and hand, and they became permanent attractions in that city's Pacific Museum of Anatomy and Natural Science. The items are believed to have vanished in the aftermath of the 1906 San Francisco earthquake and fire.

While Love and his macabre carnival traveled the countryside, the legend of Joaquín Murieta grew. In 1854, author John Rollin Ridge, a newspaper editor of Cherokee heritage living in Grass Valley, collected the various Joaquín stories and those of several other noted Californio outlaws and fused them into a single myth. From Ridge's fertile imagination sprung a book entitled *The Life and Adventures of Joaquín Murieta, the Celebrated California Bandit*. Following a common literary practice of the era, the book was published under a pseudonym: Yellow Bird, the translation of Ridge's Cherokee name, Cheesquatalawny. The introductory passage defined Joaquín's persona and set the stage for the adventures that followed:

> The first that we hear of him in the Golden State is that, in the spring of 1850, he is engaged in the honest occupation of a miner in the Stanislaus placers, then reckoned among

> the richest portions of the mines. He was then eighteen years of age, a little over the medium height, slenderly but gracefully built, and active as a young tiger. His complexion was neither very dark or very light, but clear and brilliant, and his countenance is pronounced to have been, at that time, exceedingly handsome and attractive. His large black eyes, kindling with the enthusiasm of his earnest nature, his firm and well-formed mouth, his well-shaped head from which the long, glossy, black hair hung down over his shoulders, his silvery voice full of generous utterance, and the frank and cordial bearing which distinguished him made him beloved by all with whom he came in contact.

While not shying away from the brutality associated with Joaquín, John Rollin Ridge imbued his Murieta with dash, daring, intelligence, and swagger. In one heated encounter, Ridge has Joaquín stand unafraid and unapologetic before his pursuers:

> Boyce roared out: "Boys, that fellow is Joaquín; d—n it, shoot him!" At the same instant, he himself fired but without effect.
>
> Joaquín dashed down to the creek below with headlong speed. . . . In fair view of him stood the whole company with their revolvers drawn. He dashed along that fearful trail as if he had been mounted upon a spirit-steed, shouting as he passed:
>
> "I am Joaquín! Kill me if you can!"
>
> . . . In the midst of the first firing, his hat was knocked from his head, and left his long black hair streaming behind him. He had no time to use his own pistol, but, knowing that his only chance lay in the swiftness of his sure-footed animal, he drew his keenly polished bowie-knife in proud defiance of the danger and waved it in scorn as he rode on. It was perfectly sublime to see such super-human daring and recklessness.

Joaquín was a murderer, but he had fans and admirers throughout Gold Country. Ridge made sure to highlight Joaquín's decency

"Joaquín Murieta, the Celebrated California Bandit," drawing by Charles Christian Nahl, engraving by Thomas Armstrong, *Sacramento Steamer Union,* April 22, 1855. Courtesy of the California State Library, Sacramento; California History Section.

and resolve, as when he relates the moment when Joaquín's comrade Reis kidnaps a beautiful young woman named Rosalie and injures her mother and her lover, Edward. When Joaquín learns of the abduction, he strongly rebukes Reis: "I am surprised at you. I have never done a thing of this kind. I have higher purposes in view than to torture innocent females. I would have no woman's person without her consent. . . . I ought to kill you, but since you have had some honor and manhood, I will let you off this time." Joaquín and Reis return the young woman to her mother and lover, and Rosalie demands a promise that there will not be any reprisals against Joaquín's band due to what she views as an act of kindness.

Ridge exploited common knowledge of Love's pursuit and shrewdly ended his tale with the criminal genius captured and decapitated by the relentless California Ranger. When published, the account was considered comprehensive by many. To this day, some view Ridge's story as accurate, and a few older history books still cite Ridge's fabrications as fact.

John Rollin Ridge died in 1867, but his creation did not die with him. Ridge hinted at this deathless existence when he concluded his novel with these words: "[Joaquín] leaves behind him the important lesson that there is nothing as dangerous in its consequences as *injustice* to *individuals*—whether it arises from prejudice of color or from any other source." Joaquín had become an icon, a never-ending lesson.

The creative community embraced the mythical Joaquín as an emblem of change, courage, and concern. Often, they ignored the bloodshed and violence and focused on the nobility. Artists, most notably the renowned Charles Christian Nahl, painted fictionalized "portraits" of Joaquín Murieta that were widely disseminated and sold well. In 1919, the famous silent-movie director D. W. Griffith based his only western, *Scarlet Days,* on the Joaquín Murieta legend. He passed on a young actor touted as being the perfect Joaquín—Rudolph Valentino. In 1932, Walter Noble Burns published *The Robin Hood of El Dorado,* a compilation of Joaquín stories that embellished the Joaquín-as-noble-outlier myth. In 1936, another Joaquín Murieta movie, this one based on the Burns book, was produced featuring Warner Baxter of

Cisco Kid fame. And in 1965 yet another Murieta movie arrived, this time starring the blue-eyed matinee idol Jeffrey Hunter as Joaquín.

Most folklorists believe the Zorro stories, popularized by author Johnston McCulley in his 1919 magazine serial "The Curse of Capistrano," is based at least in part on the Joaquín Murieta legend. In 1998, the two intertwined when the motion picture *The Mask of Zorro* portrayed young Alejandro Murieta (Antonio Banderas) training for and seizing the mantle of Zorro, the chivalrous servant of truth and justice, following the death of his older brother Joaquín.

As the frontier-legend and pulp-fiction roots of the Joaquín mythology receded deeper into the past, they were given new life by new generations of literary and artistic heavyweights. In 1966, playwright and Nobel Prize laureate Pablo Neruda wrote a provocative play entitled *Fulgor y Muerte de Joaquín Murieta*, or *The Splendor and Death of Joaquín Murieta*. Novelist Isabel Allende examined the Murieta myth in her 1999 novel *Daughter of Fortune*. The 1970s Sacramento art collective known as the Royal Chicano Air Force included Joaquín Murieta in a pantheon of human rights champions throughout American history, alongside Cesar Chavez, Malcolm X, and Dr. Martin Luther King, Jr. The RCAF also regarded Joaquín as a potent symbol of confrontation that aligned with its members' own political and artistic nonconformity.

The combination of Ridge's catalyst and these later creative endeavors solidified Joaquín Murieta as a leading California historical and cultural symbol. But there was more. The legend of Joaquín Murieta also became an emblem of ethnic and political resistance. Even during the Gold Rush, Joaquín Murieta metamorphosed from simply an avenging hero or bloodthirsty bandit into a figure that transcended identity. He became the symbol of justifiable retribution by the Californio community that was pushed aside by white immigrants. As early as 1853, a letter to the *New York Tribune* from its San Francisco correspondent identified only as "Geo. M.B." announced:

> [Joaquín] is one of those who welcomed Americans and American rule in California—but unfortunately one who has been despoiled over and over again, of his property;

> had his dearest rights invaded and trampled under foot by those scoundrel ruffians found in all our new settled regions, who alike disgrace our nation and a common humanity; until at length, aroused to animosity, his love turned to hate, [and] with . . . a burning revenge, he has sworn eternal warfare against everything and person American.

Mariano Guadalupe Vallejo, a powerful land baron and one of the most influential Californios, saw Murieta's actions as an understandable reaction to the rapid withdrawal of power and property from the Californios during the Gold Rush. Vallejo wrote that the "majority of the young men that had been so unjustly despoiled, burning for revenge, swelled the ranks of Joaquin Murietta and under command of this much-feared outlaw they were able to pay off some of the wrongs which the North Americans had inflicted upon them." Joaquín Murieta provides a similar release valve for many of today's Latinos; he is a daring, fearless metaphor for the disenfranchised and undocumented wishing to redress a painful, aggravating relationship with the dominant culture.

The legend of Joaquín Murieta is in parts truth, fiction, speculation, conjecture, and fabulous adventure. It reflects a mix of historical actualities, romantic visions, and cultural exasperation both past and present. As long as there are those who remember his legacy, dream of adventure, or hope for an end to discrimination, it can truly be said that "Joaquín was here."

COCK-EYE AND SNOWSHOE
JOHN CALHOUN JOHNSON AND JON TORSTEINSON-RUE, TRAILBLAZERS

John Calhoun "Cock-Eye" Johnson and Jon "Snowshoe Thompson" Torsteinson-Rue carved routes through the Sierra Nevada that led thousands of eager gold seekers into California. Both were pioneering explorers—and mailmen—but only one became legendary, while the other is largely and unjustly overlooked.

When Gold Rush society was in its infancy, mail delivery was protracted and sometimes dangerous. In the first years, argonauts usually had to visit San Francisco to send or receive mail, and the process was infuriating. Since mail could only be collected at a San Francisco post office, the lines were very long and the delay interminable. Considered that it was also a trip of sometimes hundreds of miles, it was common for mining camps to designate a courier to collect the mail. An 1854 pictorial letter sheet entitled "How We Get Out Letters" described the frustrating wait:

> Persons in the Atlantic States will readily understand the disappointment a friend here experiences at being told at the window, after undergoing for three or four hours the persecution of being jammed and jostled by an anxious crowd, exposed to a burning sun or the freezing and disagreeable winds which at seasons sweep round the corners of our streets carrying with them clouds of dust—yes, at being told by the clerk that there is nothing for him.

With the population growing rapidly, the amount of mail was staggering. Jacob Bailey Moore, the first postmaster of San Francisco, closed the post office for three days in 1849 when the facility and its five employees was flooded on a single day with "countless newspapers and 45,000 letters," as Moore recalled.

As time passed, miners came to rely on express companies to deliver their mail. These private companies ranged from one-man-with-a-mule operations to more sophisticated—or at least more audacious—enterprises. One businessman found an imaginative answer for delivering mail in deep snow: J. B. Whiting of the Feather River Express hitched teams of four crossbred Newfoundland/St. Bernard dogs to a snow sled that was capable of delivering two hundred fifty to five hundred pounds of correspondence and parcels.

But mail delivery was still routinely slow and unreliable. Harris Newmark, a Prussian immigrant, described a seemingly inevitable and violent consequence. In 1855, Newmark recalled, a criminal was sentenced to be executed, and although a stay was later granted, "the document was delayed in transit until the murderer, on January 12th, 1855, had forfeited his life!"

East of Gold Country, in the Sierra Nevada, the mail system faced even greater challenges in the earliest days of the Gold Rush. Roads were little more than hastily carved footpaths, and terrain was steep, uncharted, and buried in snow half the year. It is in this landscape that Cock-Eye Johnson and Snowshoe Thompson made their marks.

John Calhoun Johnson is the more impressive figure, but he did not receive the recognition or public exposure that he deserved. Born near Deersville in eastern Ohio in 1822, his early life story is somewhat unclear. It appears that as a young man he studied law for a short period and became a practicing attorney—a not uncommon career path at a time when lawyers often "read the law" only briefly before opening private offices. What is certain is that John Calhoun Johnson traveled to California and Nevada in the mid-1840s, and dazzled by the beauty and opportunities the region afforded, he returned home and enthralled his family and friends with stories of the wondrous sights in the faraway land to the west. Sometime around 1846, he ventured westward once again and secured the mail concession for the Lake Tahoe area.

Delivering mail from Carson City to Lake Tahoe and Placerville, he first followed the well-worn Truckee River road or the arduous Carson Pass trail. Neither was satisfactory to him: the Truckee route required crisscrossing the chilly Truckee River dozens of times, and the Carson Pass circuit was longer and crested treacherous mountain passes choked with snow most of the year. Johnson explored path after path, seeking the most efficient route. Just west of Lake Tahoe he blazed what came to be called Johnson's Cutoff, which led from the lowlands at Lake Tahoe over the summit near Echo Lake. This area is still called Johnson Pass today, and Highway 50 roughly follows its track.

Johnson's course bypassed the tricky, serpentine track of the Truckee trail, and this new way to Placerville was two thousand feet lower in elevation than Carson Pass. Johnson's path was also considerably shorter than the alternatives.

Johnson's Cutoff, later designated the Placerville Carson Valley Road, was little more than a skinny, meandering horse path in its early days, but by 1860 it had been widened and improved and was a primary route through the mountains for wagon trains, stagecoaches, and, during its brief life, the Pony Express. Within a few years of its founding, Placerville Carson Valley Road was decorated with hotels, liveries, and weigh stations. Traveling at night was preferable because the road was frequently jammed with commercial freight wagons during the day.

Near Placerville, John Calhoun Johnson, who was called Colonel Johnson or "Cock-Eye," established Six Mile Ranch, usually referred to as Johnson's Ranch. The 320-acre ranch was a welcome stopping place for Gold Rush emigrants and miners looking for new diggings. As many as one thousand visitors could camp at the spot at any one time, and there was a hotel and a general store and other outbuildings serving visitors on the ranch, as well as gold mines scattered through the property.

On July 9, 1853, the *Placerville Herald* cooed about the Fourth of July celebration held at Johnson's Ranch:

John Calhoun "Cock-Eye" Johnson: Sierra Nevada trailblazer and Gold Rush mailman. Courtesy of the California State Library, Sacramento; California History Section.

> Sweet music echoed through the vale, and soon the merry dance began, which was participated in by a very large party of ladies and gentlemen from Placerville . . . and the smaller villages of the mountains, making one of the largest and most decidedly genteel parties that ever convened in El Dorado County, upon a like occasion. . . . There was not the remotest chance for the most fastidious to complain; and in every respect was it an occasion to be remembered, with pleasure.

Johnson also managed to find time to practice law and is considered to have been one of the first practicing attorneys in California. In 1855 he was elected to the State Assembly from the Third District, and Sierra and goldfield residents found themselves in need of a new mailman. Johnson's mail route was transferred to a recent immigrant from Norway named Jon Torsteinson-Rue. Like Johnson before him, Torsteinson-Rue quickly learned that delivering a letter through the mountains in wintertime was next to impossible. But he was undeterred.

In February 1867, the *Marysville Appeal* described "snowshoes" as the "strips of board from six to ten feet long, and four inches wide, the bottoms of which are made smooth and then coated with a compound vulgarly called 'dope.'" Torsteinson-Rue strapped them to his feet in order to make his winter rounds. He and his snowshoes were laughed at, and people began to call him Snowshoe Thompson, a nickname that stuck with him the rest of his life. But once the novelty wore off, he was admired for his pluck and successful labors through snowdrifts dozens of feet high and temperatures that frequently dipped below zero. As a celebrated figure in the region, Thompson helped popularize Johnson's Placerville trail and is credited by many with having originated the route, even though in actuality he followed Johnson's already established path for most of the journey.

For many years after Johnson abandoned mail delivery in 1855, Thompson was the only mail link between the East Coast and the central portion of the goldfields. He not only carried letters but also medicine, food, and even the occasional oddity like crystal balls. According to period accounts, Snowshoe Thompson carried little food, drank snowmelt, and did not pack blankets. He slept on the open ground using his mail pouch as a pillow and a small fire to keep from being frozen solid. Snowshoe Thompson performed this demanding and hazardous task for twenty years.

Stories began to circulate about the amazing snowshoeing feats of Snowshoe Thompson. There were the tales of him "running"—or skiing in today's cross-country style—three miles in five minutes. There were the hair-raising descriptions of Snowshoe leaping off a ninety-foot precipice and landing without a scratch. Some of these accounts of derring-do were accurate, while others were just tall tales.

Today we remember Snowshoe Thompson, while his predecessor, Cock-Eye Johnson, inhabits the historical shadows primarily due to lacking the one major advantage Thompson possessed: Snowshoe had an admiring chronicler who ushered him into myth. His name was William Wright, a reporter for the *Virginia City Territorial Enterprise*, who used the pseudonym "Dan DeQuille."

DeQuille wrote that Thompson was "adventurous, fearless, and unconquerable" and deserved a "hero's crown." DeQuille was nothing less than effusive about Snowshoe. Consider this feverish passage from an 1886 *Overland Monthly* article:

> To ordinary men there is something terrible in the wild winter storms that often sweep through the Sierras; but the louder the howlings of the gale rose, the higher rose the courage of Show-shoe Thompson. . . . In the turmoil of the most fearful tempests that ever beat against the granite walls of the High Sierras he was undismayed. In the midst of the midnight hurricane, he danced on the rocks as though himself one of the genii of the storm.

Dan DeQuille solidified the mythology of Snowshoe Thompson, but his celebrity was already hardening when DeQuille put his quill to paper. Not everyone admired "the genii of the storm" or viewed Thompson as the exemplar of snowshoeing skill, and sometimes this led to heated words and fierce accusations of fraud, contentions that ironically may have helped his legend grow. Today, Snowshoe Thompson is regarded as one of the pioneers of modern skiing in the United States. He has been memorialized in many spots in the Sierra Nevada with plaques, stone markers, bronze statues, and an elaborate monument in Genoa, Nevada. Meanwhile, John Calhoun Johnson has steadily receded into the background.

But Cock-Eye was not forgotten by those who knew him best: the residents of Placerville and El Dorado County. In 1883, years after both Johnson and Torsteinson-Rue had died, *The Historical Souvenir of El Dorado County,* a county history profiling its "prominent men and pioneers," wrote:

> While we do not wish to depreciate the services and merits due to [Snowshoe] Thompson, it is due to truth and justice also to state, that one of the earliest settlers of El Dorado county, Jack C. Johnson, of Johnson's ranch, preceded Thompson as a trans-mountain mail carrier; he was the man that opened up, marked out, and traversed the route called after him, "Johnson's Cut-off," which subsequently was traveled by Thompson, and he crossed the mountain range through the depression laid down on all the maps as "Johnson's Pass." . . . It is not more than right that the government appreciated Thompson's services who intrepid and faithfully did his difficult and dangerous duty, unconcerned of season and weather, but let the truth of history be vindicated. Jack Johnson claims the name as the . . . pioneer of trans-mountain mail carrying on foot by the Placerville route.

DUELING DESIGNATIONS

LAKE BIGLER VERSUS LAKE TAHOE

The argonauts focused their attention on the western Sierra Nevada foothills, the fabled Mother Lode, but some mining also occurred in the eastern Sierra Nevada. Most miners, however, avoided the area due to severe weather, difficult terrain, or the still-fresh memories of the tragic fate of the Donner Party. Wagon trains generally viewed the Sierra Nevada crest an obstacle, the final hardship on their months-long journey to the land of dreams. But for some, it was a land of dreams itself; many travelers and a handful of miners remembered the extraordinary scenery of the Sierra high country and yearned to know more, to experience more. As the whirlwind of activity swept through Gold Country, transforming the social and economic landscape, there were a few who found a different type of treasure in a stunning sheet of crystal-clear water high in the Sierra Nevada just south of Truckee and Donner Lake. Even more had heard rumors of this scene of unequalled enchantment and hoped to visit someday. The site was Lake Tahoe, as we all know it today.

David Augustus Shaw, an 1850 gold seeker from Illinois, gave voice to the adulation of those who had encountered the "rich, wild, but picturesque and beautiful scenery" of the splendid mountain lake.

> The soul catches that sweet inspiration which calmly draws us into communion with the harmony of nature and a contemplation of a better land as we stand upon the silvery

> shores of Lake Tahoe, while amidst a stillness sublime and awful the rays of the morning sun, like ribbons of gold dart through the dense forest, streaking with amber and golden sheen the dark blue waters through whose transparent depths the landscape is mirrored below, God's fountain in the wilderness to beautify his footstool and invigorate all his creatures that partake of its crystal waters.

There was never any doubt as to the beauty and soul-nourishing power of the lake, but there was a question as to what to call it. In the Gold Rush years and afterward the appellation "Lake Tahoe" was far from certain or official. There was a case of dueling designations.

On crisp and snowy February 14, 1844, an expedition led by John C. Frémont, "The Great Pathfinder," crested a ridge and looked north at a breathtaking panorama: "We had a beautiful view of a mountain lake at our feet, about 15 miles in length, and so entirely surrounded by mountains that we could not discover an outlet." Fremont's party is considered the first non-Native group to have seen Lake Tahoe. But the title "Tahoe" was not assigned to the maps that were produced in the years immediately following Fremont's "discovery." Charles Preuss, Fremont's cartographer, simply called it "Mountain Lake," and that label was commonly used to designate the site until 1852. Fremont himself named it "Lake Bonpland," after Aimé Jacques Alexandre Bonpland, the French botanist who had accompanied Baron von Humboldt on his Western Hemisphere explorations in the early nineteenth century, and that became the preferred name of European mapmakers.

Some commercial maps sold to early gold seekers identified the lake as "Fremont's Lake"—if the lake was shown at all. On an 1853 map, William M. Eddy, the surveyor general of California, styled it "Lake Bigler," in honor of newly elected California governor John Bigler, who, in 1852, had led a party that rescued a snowbound wagon company south of the lake. It is believed that John Calhoun Johnson, the pioneering Sierra Nevada trailblazer and mailman, was the first to suggest the name "Lake Bigler."

Other Gold Rush–era maps and guides used other names, such as "Big Truckee Lake" or the hybrid "Lake Bigler Tahoe." And just to confuse things even further, George Holbrook Baker's popular 1855 "Map of the Mining Regions of California" used several of the names in contention to refer to different lakes on the same map: "Mountain Lake," "Lake Bigler," and, at the location where Lake Tahoe should be, "Mahlon Lake."

Of all the potential choices, it was "Lake Bigler" that stuck, at least for a while. When the Civil War began in 1861 and former governor John Bigler became a controversial figure—for his support of the Democratic Party, for his Southern, pro-slavery sympathies, and for his possible involvement in a secession plot in California—supporters of the Union cause called for the name "Lake Bigler" to be stripped from the beautiful alpine lake. Union adherents even introduced legislation in the 1861 California State Assembly to rechristen the lake "Tula Tulia," which a handful claimed was the appropriate Native designation.

In 1862, according to his personal papers, Robert Dean, who managed the Lake House resort on the lakeshore, wrote to the Union-sympathizing *Sacramento Union* suggesting a name change. Dean wrote that he was acquainted with a local Washoe Indian leader known as

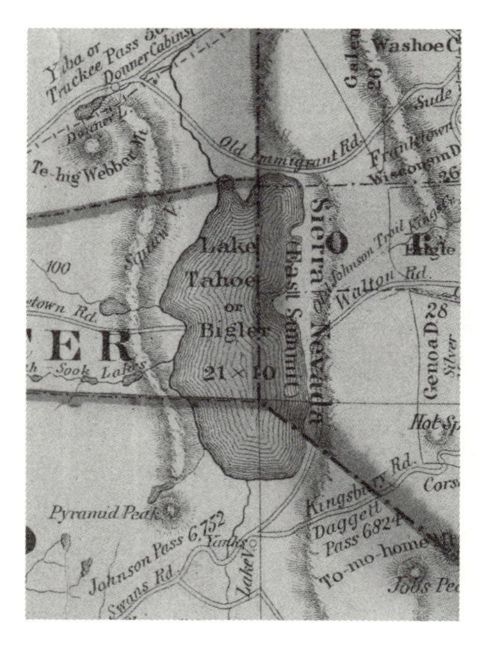

For decades, there was a lively debate as to the proper name of Lake Tahoe. During the Gold Rush, the argument centered on "Lake Tahoe" versus "Lake Bigler." Detail from "DeGroot's Map of Nevada Territory," 1863. Courtesy of the California State Library, Sacramento; California History Section.

Captain Jim, who had informed him that the Natives referred to the lake as "Tahoo." In May 1863, the *Union* published an editorial supporting the name change to basically anything other than Lake Bigler. It suggested, "Sierra Lake would be very well, or the Indian name of the lake, Teho," which was the newspaper's rendering of the local Indian name suggested by Robert Dean.

In June 1863, the Territory of Nevada entered the debate. The *Nevada Transcript* wrote: "It is proposed to drop the name of Bigler from the lake among the Sierras and adopt the Indian name of Tahoe. Good idea. Why the finest sheet of water in the mountains should be named after a fifth rate politician we have never been able to see." Another Nevadan proposed the name "Lake Union."

Critics of John Bigler pounced. Many referred to Bigler's fondness for and frequent overconsumption of lager beer as reason enough to change the lake's name, and comments often referenced the widely reported fact that Bigler carried a packet of powered soda and magnesia in his vest pocket to calm his stomach after his epic drinking bouts. The *Marysville Appeal* bitingly suggested that Lake Bigler be renamed "Lago Beergler," as it "would stand always as a punning allusion to the bibulous habits of 'Honest John' when he was Governor of the State." The *Sacramento Union* responded to the *Appeal*'s proposition, saying that it objected to both Lake Bigler and Lago Beergler unless, in fact, "a lake of beer should be discovered." The *Union* then offered their simple solution to the Bigler-Tahoe conundrum on July 27, 1863: "[Bigler should] apply to the next Legislature for leave to call himself John Tahoe."

In summer 1863, a renowned Unitarian preacher, the Reverend Thomas Starr King, weighed in. Tahoe, he wrote, "stems from the Washoe Indian 'Tache' [meaning "much water"] plus 'Dao' [meaning "deep or blue water"] and is the name that should rest upon its water." But Mark Twain, then writing for the *Virginia City Territorial Enterprise*, strenuously disagreed with the lake being renamed Tahoe, which he called a "disgustingly sick and silly . . . name."

While there was ongoing support for Lake Bigler in some circles, it was falling out of favor, and in 1870, there was yet another twist. After the Civil War, the California Republican Party was in disarray and the

Democrats regained control of the state legislature. On February 10, 1870, the now ascendant Democrats stuck the knife to the defeated Republicans by overwhelmingly passing a new law, Chapter 58 of the Statutes of California, which stated, "The lake shall be known as Lake Bigler . . . the only name to be regarded as legal." And yet despite the formal sanction of Lake Bigler in 1870, that moniker was by then rarely used; Lake Tahoe had come to be accepted far and wide.

The next questions became how do you spell and pronounce "Tahoe"? Many spelling variants were proposed over the decades: "Tah-ve," "Tah-oo-e," "Tah-ho-ee," "Tahoo," "Tah-jo," "Ta-ho," "Tajo," "Pah-hoe," "Ta-au," and "Tahoe." Pronunciations included "Tah-hoe," "Tay-hoe," "Daw-oh," and even "Tah-joe." This led the *Truckee Republican* to publish, in 1886, a whimsical primer entitled "'Tahoe' as She Is Spoke," recalling a conversation among seven Lake Tahoe tourists and a *Republican* reporter:

> They were discussing the beauties of our grand mountain lake.
>
> First tourist—I think Lake Taw-who is simply magnificent.
>
> Second tourist—Tay-o's sunsets are especially gorgeous.
>
> Third tourist—And the water; I never saw such a beautiful shade of blue colors as in Lake T'-hoo.
>
> Fourth tourist—They say that the body of a person drowned in Lake Tay-how never rises.
>
> Fifth tourist—I never ate finer trout than are caught in Lake Taw-o.
>
> Sixth tourist—I am going to fetch my aunt and mother-in-law up to *Tay*-ho next summer.
>
> Seventh tourist—I declare I mustn't miss going to Ta-ho [short "a"] to-morrow.
>
> Just then the reporter remarked that *Tah*-hoe was attracting larger crowds of pleasure-seekers every year.

The lake remained officially named "Lake Bigler" until July 18, 1945, when California Senate Bill 1265 repealed the 1870 act and authorized "Lake Tahoe."

"A GREAT DEAL OF HARD TIMES"

THE DIARY OF JOSEPH PIKE

It is a mountain of words.

Happening within a literate society that valued letter-writing, journal-keeping, and use of the "information highway" of the period—newspaper correspondence—the Gold Rush has been documented in hundreds of accounts. The eyewitnesses were farmers, merchants, and city dwellers far more comfortable wielding a pen than a gold pan. Frequently their accounts begin with boundless enthusiasm and ardent anticipation but become terser and more serious as the exacting realities of the experience mount.

This Everest of words can be maddening. Many journals are a formal, tedious slog through a company ledger book, and some accounts are regularly filled with cringe-inducing sentimental poetry or strained attempts at lyricism that employ fifteen words where three would have been sufficient. Occasionally, the chronicles are merely sleep-inducing registers of miles traveled or temperature readings. Letters can suddenly veer from fascinating tales to pompous, judgmental diatribes. Often we long for more details than we get. Was he tall? Why did she suddenly stop writing? Did the dead miner have a family? How did he break his leg? What happened to that grizzly bear?

But then there are the exceptions: lively accounts of a world transformed, soul-stirring narratives of danger and personal discovery, wonderful encounters with memorable eccentrics, joyful tales, and poignant sketches of loneliness and loss. Some of these chronicles are justifiably well known, such as Dame Shirley's letters, crackling with life, that she sent to her sister back home. Others are more obscure but just as entertaining and enlightening, such as the ribald diary of Alfred Doten or the acerbic commentary of Scottish artist John David Borthwick. There are literally hundreds of unpublished diaries and journals that shed light on this extraordinary moment, and occasionally we find a gem—a hidden, homespun tale that details the day-to-day experiences of a gold seeker flush with hope but realistic to his core. The text may not be flowery, the details can be sketchy, but the story is honest and heartfelt. Such a diary belonged to Joseph Pike of Half Day, Illinois, a village about thirty miles north of Chicago.

As is often the case in these diaries, we know few specifics about its author. The diary covers twenty months, from April 1850 to December 1851, including Pike's stay in the California goldfields, from September 1850 to October 1851. Upon his arrival in Gold Country on September 19, 1850, Pike wrote:

> Today about two oclock we landed in Georgetown amid the gold diging of California and found the long sought for elderado so long looked for after traveling over 2500 miles of the most desolate part of the whole globe for nearly two thousand miles of the distance and seen more misery than I ever saw in my life before by ten times multiplied by ten. Have got my fill for once of traveling although I consider myself tolerably well paid for my panes and if I ever get any gold I do not consider myself all lost for it is a great school for the man that wants to learn.

Joseph Pike set to work immediately, plunging headlong into placer mining near Cold Springs, halfway between Coloma and Placerville. To his delight, he found success. Pike wrote on September 28, 1850, "Returned home in the morning and washed out a hole got about $150, for 7 days work Shurely thought we had found the Eliphant." One can understand his excitement, as $150 in 1850 would be worth $4,550 today.

Pike continued to strike paydirt. He meticulously recorded his gold mining results for November 1850 in his elaborate penmanship:

November	
1	$42.40
2	$25.60
5	$21.00
6	$65.50
7	$39.00
8	$12.00
9	$53.00
11	$39.00
12	$29.80
15	$38.00
16	$45.00
18	$32.00
19	$25.00
21	$67.50
28	$18.00

His November haul would be worth $16,751.52 today.

But in Gold Rush California, that $50 gold strike might only be just enough to pay for that night's dinner. It was pure supply-and-demand economics; supplies were often extremely low, and prices inflated dramatically. James Carson, an early gold seeker, described the startling rise in the purchase price of stock animals:

> In 1846 and '47, the price of the finest horses was $20; fat bullocks, $6; wild mares, 75 cents each. . . . The discovery of gold raised the price of stock in proportion with everything else. Horses and mules in the mines were worth from two to four hundred dollars; cattle from one to two hundred dollars per head. I have seen men give two and three hundred dollars for mules and horses—ride them from one digging to another—take their saddles off, and set the animals loose, (never looking for them again,) remarking that 'it was easier to dig out the price of another, than to hunt up the one astray.'

Other costs were equally astronomical:

Item	1850 price	2016 price
1 lb. butter	$20.00	$606.00
1 lb. cheese	$25.00	$758.00
1 egg	$3.00	$91.00
50 lbs. flour	$13.00	$394.00
Pair of boots	$6.00	$182.00
Flannel shirt	$1.50	$45.00
Blanket	$5.00	$157.00
Shovel	$36.00	$1,090.00

Just to make ends meet, the hard work had to be unceasing, even on Christmas Day. Joseph Pike described his 1850 holiday: "This [day] was quite cold and frosty but very mild and quite dry. Washed out $39.00 dollars. This has been a day of uproar and confusion[;] well that Christmas does not come but once a year."

As the drudgery continued for month after grueling month, Pike's diary entries became shorter. His few words revealed his exhaustion and disillusionment. Yes, Pike found gold, but it was not enough to be a profit and never enough to justify a return home. He was sore, tired, and frustrated. He wrote on September 3, 1851: "[These were] two as hard days work as I have done lately. Oh this making a mule of ones self he might as well be a jackass and be done with it."

After a year in the diggings, Joseph Pike reflected on his experiences, both good and bad. On September 19, 1851, his diary read: "Today I have been here in the mines one year have enjoyed life in various forms seen a great deal of hard times as well as considerable prosperity and good times if a life in this dreary region can be called good."

The next day, Pike was seriously hurt in a fall. A week later, still racked with pain—"my heart paned me much," he wrote—he left his home of one year, having decided to "take a tramp south and see what I could find." What Joseph Pike found was a desire to return to Illinois. On October 6, 1851, he described his unpretentious departure: "Returnd to Cold spring and went to Hangtown saw Norman Stevens and other acquaintances bid them good by."

110
15 To day dug and washed out $ 14.50
To day Mr [illegible] came to tent with us till
such a time as he and Mr [illegible] start home
16 To day dug and washed out $ 12 50
To day noon Mr [illegible] began to work with us.
17 To day dug and washed out $ 17 00
some signs of rain but cleared up, cool nights
18 To day worked back towards the ravine got $22, 00
19 To day worked back nearer the ravine got $22 50
To day I have been one year in the mines have enjoyed
life in various forms seen a great deal of hard times
as well as considerable prosperity and good times
if a life in this dreary region can be called good
20 This morning while getting a back stick my feet
slipped on a smooth rock and I fell and hurt me so
bad that I gave up work for the day, mended
pants and went to look for diggings
21 Sabbath A windy day otherwise pleasant
spent it alone and as usual in reading and
thinking When will I enjoy the privileges of
my family and society in a christian land
22 To day we dug and washed out $ 11.25
23 To day dug and washed out $ 15.50
24 In the fore noon got out $ 4 00
Sept. 19, 1851

111
In the after noon moved our tom and got ready
for another day. My lameness very bad
25 Last night my breast pained me much and I
made up my mind that I had got to quit work for
a while and as my diggings were nearly worked out
I thought I would take a tramp south and see what I
could find, got ready and started about ten oclock
crost Bear river at the [illegible] crossing through Illinois
town then down [illegible] vale and Rabit dale to the
North fork of the American river put up for the night
[illegible] tired and weary, traveled some fifteen miles
26 This morning took a zig zag course up the mou
ntain out of the north fork valley for over a
mile then into Indian Canon called on some
old chums and took diner then called on Mr
Archer at Elizabethtown past on through
Shirttail Canon to Yankee Jims diggings and
put up for the night at Mr [illegible] traveled 12 m
27 This started for Georgetown Traveled 20 m
Sabbath 28 Spent the day with Mr Smith
saw Mr John [illegible] or Uncl John
29 Started early this morning for [illegible]
Cold Spring by the way of Coloma, stoped

The diary of Joseph Pike, an Illinois gold seeker, is one of the simplest, and most edifying, of all Gold Rush journals. Pages from *Diary of a Forty-niner, Being the Diary of Joseph Pike of Half Day, Illinois, April 15, 1850, to December 29, 1851.* Courtesy of the California State Library, Sacramento; California History Section.

Joseph Pike had been to California. He had "seen the elephant." Now it was time to go home.

GUILTY OF DUST AND SIN

DRINKING, DINING, AND DESIRE

Once a week, Porter Rockwell, the "Destroying Angel of Mormondom," would arrive on a hill above Buckner's Bar on the Middle Fork of the American River with a string of mules packing whiskey. Upon reaching the crest of the ridge, Rockwell would unload his cargo and blow a horn to announce his presence. Jack Smith, Rockwell's confederate in the teeming mining camp down below, would then fire his pistol to herald that Old Port's whiskey store was open for business. According to the 1882 *History of Placer County* by publisher Thompson and West, miners would scramble up the hill to partake of Rockwell's liquor to get "gloriously drunk." Rockwell may have been Brigham Young's personal bodyguard, but he also knew a good business opportunity when he found one.

In the earliest years of the California Gold Rush, all was impermanent, free, and unregulated. There was very little law, there were no immediate family responsibilities, and there were no impediments to brazen immoderation. Success would bring celebration, failure sorrowfulness, and often these would be exhibited in identical forms of abandon. From the mildest to the wildest manifestations, Gold Rushers felt unrestricted by societal conventions and precepts. From getting "gloriously drunk," gorging themselves like "wolves or anacondas," indulging in sexual "allurements still harder to resist," or amusing themselves with new technology, gold seekers were liberated from the expectations of genteel culture. And they took full advantage.

With the possible exception of gambling, with which it often went hand in hand, drinking was the favored pastime of the California Gold Rush. Virtually every commentator, journalist, and diarist confirmed that the consumption of spirits was widespread. In 1850, English adventurer Frank Marryat marveled, "Drinking is carried on to an incredible extent here; . . . a vast quantity of liquor is daily consumed." Leonard Kip, a forty-niner from New York, remarked that, during his two months in the goldfields, he encountered camps every day where appetites of all sorts were augmented by alcohol: "Wines and brandy flow freely, dice are brought out and particularly monte enchains eager groups around the different tables; lotteries are in full blast; occasional fights arise; and, on all sides, commences a scene of riot, drunkenness and wrangling. . . . Whatever the occupation, strong drink flows freely, and oaths and coarse songs hold a large place in the revels."

Many chroniclers discerned that a trait of California society was the degree of conviviality and comradeship exhibited during these drinking bouts. The acerbic Gold Rush Scotsman John David Borthwick felt this reflected the participants' shared experience and a sense of urgency:

> [These young men] make the voyage through life under a full head of steam all the time; they live more in a given time than other people, and naturally have recourse to constant stimulants to make up for the want of intervals of *abandon* and repose. . . . The bars are the most favourite resort, being situated in the most frequented and conspicuous places; and . . . at all hours of the day, men are gulping down fiery mouthfuls of brandy or gin. . . . No one ever thinks of drinking at a bar alone: he looks round for some friend whom he can ask to join him; it is not etiquette to refuse, and it is expected that the civility will be returned.

Drinking and the camaraderie it could engender were a means to escape loneliness or homesickness, or to drown the bitter taste of failure. Too often the result was not temporary relief but permanent infirmity. As John David Borthwick observed,

> Drinking was the great consolation for those who had not moral strength to bear up under their disappointments. Some men gradually obscured their intellects by increased habits of drinking, and, equally gradually, reached the lowest stage of misery and want; while others went at it with more force, and drank themselves into *delirium tremens* before they knew where they were. This is a very common disease in California: there is something in the climate which superinduces it with less provocation than in other countries.

Any occasion could and would be used as an excuse to indulge, but holidays were a readymade reason for celebration. Dame Shirley, the pen name of Louise Amelia Knapp Smith Clappe, wrote a series of letters about life in the Gold Rush community of Rich Bar in the Feather River Gorge, including the "Saturnalia" of Christmas 1851 at the Humboldt Saloon. Dame Shirley described the long line of mules delivering "casks of brandy and baskets of champagne" to the establishment and the "oyster-and-champagne supper" that followed. Once the liquor began flowing and the feast was consumed, dancing began. As she recalled, "I believe that the company danced all night. At any rate, they were dancing when I went to sleep, and they were dancing when I woke the next morning. The revel kept up in this mad way for three days, growing wilder every hour." On the fourth day, Dame Shirley reported, "they got past dancing, and, lying in drunken heaps about the barroom, commenced a most unearthly howling. Some barked like dogs, some roared like bulls, and others hissed like serpents and geese. Many were too far gone to imitate anything but their own animalized selves."

Alfred Doten, renowned as a foremost "reveler," described Christmas 1853 in his diary. Doten found himself in Fort John, Amador County, where he threw a "Christmas Spree" featuring "a glorious game supper of fried deer tongue, liver, quails, and hares," washed downed with barrels of cognac and accompanied by violin, flute, banjo, clarinet, and accordion music and numerous gunshots. A dance followed, and Doten and his fellow merrymakers did the Highland Fling and

Next to finding gold, among the forty-niners' favorite activities were drinking, dining, and other exuberantly unsavory amusements. Image 7 from *The Miners' Pioneer Ten Commandments of 1849.* Courtesy of the California State Library, Sacramento; California History Section.

---

a series of what he described as "Variations" on an improvised dance called the "Double-Cowtird-Smasher."

At times, the revelry became so rowdy that measures were needed to dull the potential risks. Arrests, dispersing a drunken throng, or escorting a staggering, threatening carouser into the woods to sleep it off were familiar solutions. Bayard Taylor, a twenty-four-year-old Gold Rush correspondent for the *New York Tribune*, remembered one intervention on behalf of a drunken gambler: "The man's friends took away his money and deposited it in the hands of the Alcalde, then tied him to a tree where they left him till he became sober."

The consequences of a night of overindulgence could be more serious than bar brawls or fortunes lost at a gambling table. William Perkins, a Sonora merchant, recalled, "The most common and fatal result . . . of drunkenness is falling into some of the thousands of deep pits, dug during the summer by the miners, and now full of water. Scarcely a week passes that two or three bodies be not fished out of these holes."

John David Borthwick also recorded that drinking was inextricably linked to another form of overindulgence: overeating. Proprietors of "grogshops" found that the competition for customers was nearly a contact sport and that offering "merely a plate of crackers and cheese on the counter" was not enough. Saloonkeepers began to daily provide a table "covered with a most sumptuous lunch of soups, cold meats, fish, and so on,—with two or three waiters to attend to it. This was all free—there was nothing to pay for it."

Having a meal during the Gold Rush often became an adventurous mixture of essentials and excesses. For many of the young male participants, cooking was a foreign concept, and in the goldfields they often just made do, surviving on beans, bacon, "burnt offerings," and dreams of finer feasts once they struck paydirt. Restaurants were crude affairs in the diggings, often featuring a table that was merely a wooden plank stretched between two barrels. The quality of the food and drink could vary dramatically, but the meals always possessed the highly desirable attributes of being cooked by someone else, who would also do the dishes.

Dining in the cities could be a raucous affair characterized by unmannerly speed and atrocious etiquette. While superior restaurants offering gourmet cuisine and quality service existed, the majority of gold seekers desired quick meals served in venues best described as pig trough buffets. The 1855 *Annals of San Francisco* declared, "Time was too precious to stay indoors and cook victuals. Consequently, an immense majority of the people took their meals at restaurants, boarding-houses and hotels. . . . Many of these were miserable hovels, which showed only bad fare and worst attendance, dirt, discomfort and high prices."

In 1851, Englishman William Shaw described a typical Gold Rush "eating-house" and the mealtime feeding frenzy:

> The eating-houses are peculiarly Californian in character; they are long plank buildings in the shape of a booth, having two rows of tables, placed parallel to each other, extending the length of the room. . . . At certain hours in the day, the beating of gongs and ringing of bells from all quarters, announce feeding time at the various refectories; at this signal a rush is made to the tables. It is not uncommon to see your neighbour coolly abstract a quid [of tobacco] from his jaw, placing it for the time being in his waistcoat pocket, or hat, or sometimes beside his plate, even; then commences, on all sides, a fierce attack on the eatables, and the contents of the dishes rapidly disappear. Lucky is the man who has a quick eye and a long arm; for every one helps himself indiscriminately, and attention is seldom paid to any request.

Genteel manners in these rough-and-tumble enterprises were but a rumor. In an 1850 journal entry, Dr. Israel Lord, a stiff, respectable physician from New England seeking his fortune in California, was astounded by the "utter recklessness" of eating habits, the "worst fault" of the Gold Rush population: "They gorge themselves on beef & bread and stale butter and peppered pickles & stimulating sauces, like wolves or anacondas." William Shaw commented that the "less refined" diners

> neither use fork or spoon, the knife serving to convey to the mouth both liquids and solids, which is done with surprising velocity. The voracity with which they feed is equal to the rapidity of their movements; ten minutes being the usual time for dinner, frequently less. . . . Dinner being over, the table is replenished for a second party; whilst the greasy knives are wiped, preparatory to being replaced, it is not unusual to see one of the satiated picking his teeth with a fork.

These frantic benders, unappetizing meals, and deplorable manners could only be the result of one yawning Gold Rush shortage, wrote the grumpy North Carolinian critic Hinton Rowan Helper; according to him, the "very important cause of this wild excitement, degeneracy, dissipation, and deplorable condition of affairs, may be found in the disproportion of the sexes—in the scarcity of women." And, Helper implies, the lack of sexual release.

Sex is infrequently mentioned in Gold Rush literature, but that doesn't mean there wasn't any. Sex was obviously not a subject matter routinely included in letters home to Mom, Dad, or a patiently waiting spouse, but in letters, diaries, and journals that were never expected to be analyzed or published it's a different story.

Sexual allusions were sometimes blended into descriptions of Gold Rush drinking, dining, and gambling establishments, part and parcel of the same multifaceted and phantasmagorical complex. In an October 1853 letter, one anonymous gold seeker reckoned that carnal desire began in the gambling halls where "the 'Bankers' . . . are generally women chosen for their *attractive* powers—and you will not wonder that the poor Devil who has been so long away from civilization becomes reckless, and forgits in the excitement, every thing but the Present." If the "poor Devil" manages to avoid this initial enticement, the neighboring business features a front room with

> a 'Bar' which is tended by girls, who will . . . then invite you into the parlor whose doors are always open—here you will find girls dressed in the most magnificent apparel, dancing, walzing, Playing the Piano, or guitar . . . lounging on the sofa, and sipping her wine . . . at the expense of the poor Devel, on whose shoulder she leans, until he, half Drunk, or crazy, follows her to chambers and awakes the next morning, to find himself worse off perhaps, than the one who got "fleeced" at the other house.

Occasionally, a man would begrudgingly admit that he (or, "someone I knew") had succumbed to the potent lure of desire. In a moment of guilt-driven candor, Henry Packer, a forty-niner from Pittsburgh, Pennsylvania, and a onetime Quaker, wrote to his fiancée, Mary Elizabeth, in 1853, "Sometimes I only wish you were here to watch and guard me, for I am horably beset with temptations, and though my virtue is of the most stubern kind—yet, unaided and unsustained by kindred virtue it might—I say might fall." Henry then confesses that he had visited a brothel, shamefacedly adding, "I did go in just once—only once, and then but for a few minutes." Packer explained that in a moment of weakness, he noticed a "fairy form."

> By Heaven a woman stands in the door—she is richly dressed. In her ears and on her fingers are massive gold rings displayed around her neck is a chain of the same. Glossy curls play over her full neck and shoulders. On her countenance plays a smile that would bewitch if not be guile a minister. . . . She speaks. "Come in you fellow with the mud on your hat, I like a miner."

Henry Packer begged Mary Elizabeth for forgiveness. We have no record of her response.

A handful of gold seekers were unapologetic about their sexual cravings. Never expecting anyone but his friends and companions to know of his activities, Simon Stevens, a young miner in Woods Diggings, near Sonora, wrote to his cousin in Rockland, Maine, in 1853:

> You spoke in your letter about all of you being married[;] your all getting a damd lot of good fucking; suppose you think you have got the advantage of me but i got a crack at a very hansom girl last sunday it only cos me a 20 fucking is from 5 dollars out here.

Desire can take many forms, varying from outrageous to ordinary, but of course not every member of the Gold Rush community indulged in the carnival of earthly enticements. Many gold seekers perceived their task as a once-in-a-lifetime opportunity, a solemn mission to forge their destiny, and others were obligated to sponsors back home, and for these men, there was no time for frivolous affairs or wasted days. While the excesses of others might garner the most attention, the desires of serious, sober-minded argonauts were often unpretentious, such as a good meal or a peaceful night's sleep. A frequent yearning was to permanently capture a memory of their adventure or to reassure friends and loved ones back home that all was well. For many, this was satisfied by a heartfelt letter home, but for a growing number, that keepsake or reassurance came in the form of a relatively new technology: the photograph. The Gold Rush was the first worldwide event to coincide with this significant advance in personal communication; no longer limited to drawings, etchings, or paintings, we now see actual human faces attached to a historical incident.

The first practical photographic process was introduced less than a decade before the Gold Rush began. In 1839, French inventor Louis Daguerre launched the eponymous daguerreotype process. A silver-plated copper sheet was polished, treated with fumes to make the surface light-sensitive, and then exposed in a camera. An image would appear after a second fuming with mercury vapor, and the resulting plate was then rinsed and dried. The process required long exposures, and the subjects of the daguerreotypes had to remain perfectly still to avoid the image coming out blurry. In the resulting pictures, subjects often looked exceedingly uncomfortable and stiff; painful grimaces were more common than smiles. But when done right, the pictures were astonishingly sharp and detailed. Daguerreotypes were so delicate and easily damaged that they had to remain sealed behind glass inside protective cases slightly larger than a deck of cards.

By the advent of the Gold Rush, daguerreotypes were very popular throughout Europe and the United States. As Gary F. Kurutz, the curator of Special Collections for the California State Library in Sacramento, observed, "Those one-of-a-kind, silvery, mirror-like images . . . provide a breathtaking, crystal-clear view of life during that rambunctious era."

For the largely male, mostly young Gold Rush emigrants, carrying a daguerreotype of sweethearts, wives, daughters, or other family members was *de rigueur*. Upon arrival in the land of golden dreams, a common rite of passage was to visit a daguerreotype parlor and record the momentous occasion. In 1850, Theodore Augustus Barry and Benjamin Patten, two forty-niners from Massachusetts who operated a saloon in San Francisco, discussed the common rationale for the images:

> [These new] Californians were so anxious that their friends in civilized countries should see just how they looked in their mining dress, with their terrible revolver, the handle protruding menacingly from the holster, somehow, twisted in front, when sitting for a daguerreotype to send 'to the States.' They were proud of their curling moustaches and flowing beards; their bandit-looking *sombreros*.

The desire to get "daguerreotyped" spawned dozens of studios, from the larger cities to the smallest mining camps throughout Gold Country. An 1851 advertisement for the "Daguerrian" duo of J. Lewis and George Reuben trumpeted: "A good opportunity is now offered to them who wish to give their friends in the States an idea of the effect which a rough and tumble life, and a California sky and climate has had upon their personal appearance."

PLAYERS AND PAINTED STAGES

CURIOSITIES AND AMUSEMENTS

It was April 1853. James Ayers, a forty-niner from Missouri, was peering anxiously out his hotel window in the El Dorado County community then known as Mud Springs, a few miles west of Placerville. Today known as El Dorado, Mud Springs was once a prosperous mining camp with a considerable population. Ayers had led a varied life since his arrival in Gold Country four years earlier. He unsuccessfully tried his hand at mining in Calaveras County, he was a journalist in Mokelumne Hill for two years, and now he was part of a two-man theatrical troupe plying the goldfields. This night, Ayers and his partner were anticipating a large, enthusiastic audience for their performance of comedic sketches and songs. In expectation, the troupe had rented a spacious hall and engaged an "orchestra" composed of a violin and two accordions. Adding room and board to their expenses, the duo was broke, but they expected to turn a profit with a full house that evening. Ayers was concerned as he looked to the sky from his window. It was raining, but he figured it was merely a passing April shower. James Ayers was wrong.

As the thespians repaired to the theater to make preparations for that evening's performance, the rain continued and increased in violence. When they opened the doors, standing in line was one lone miner. He took his seat and waited for the curtain to rise. Ayers described how they could hear the rain "beating upon the roof like a shower of bullets." The storm became a deluge. No other patrons arrived and it was time to begin.

Ayers and his companion decided to ask the single audience member if he wanted his money back. But, Ayers wrote, "I found our solitary individual a good-natured fellow, and he seemed to be delighted with the idea that he could have the performance all to himself. I told the two accordeons and the fiddle to tune up and we went on the stage to dress." Ayers himself found the experience novel and wonderful: "I had never before played to a more enthusiastic audience. The man was delighted with everything we said or did—guffawed outright at every [word]." After the curtain fell, Ayers recalled, "I asked him how he liked the performance. He declared it was the best he had ever seen, and said he would come every night. When we started for the hotel, I took the whole audience out and treated it, notwithstanding to do so consumed all the money that had come into the treasury."

Gold Rush entertainment was a cherished respite from the drudgery and disappointment of mining, the aching backs, homesickness, and loneliness. The range of entertainment was impressive. There were the simple pleasures of a fire, a pipe, and a tall tale, and there were also elaborate spectacles, circuses, music halls, and concerts that charmed hundreds. There were impromptu homemade, handmade frivolities alongside celebrity performers fresh from the stages of New York, London, and Paris. There was good opera, bad Shakespeare, titillating melodrama, energetic minstrels, and talentless dancers. These entertainments were often audience-participation extravaganzas, whether the performers wanted it or not.

Amusement became an integral element in the mining camps. A case in point is the Calaveras County town of Mokelumne Hill, which had grown rapidly after gold was discovered there in 1848. Centrally located in Gold Country and with a good road leading from Stockton and advantageous placement along a route that featured a two-thousand-foot-tall pyramid-shaped hill that served as a natural guidepost, Mokelumne Hill provided easy access to Placerville and Jackson to the north and Angels Camp and Sonora to the south. Chosen as the Calaveras County seat in 1852, Mokelumne Hill boomed, transforming from a haphazard sprinkling of canvas tents into a thriving commercial crossroads.

In construction, Mokelumne Hill was a typical central–Gold Country mining camp. Observer John David Borthwick noted that "the town itself, with the exception of two or three wooden stores and gambling saloons, was all of canvass. Many of the houses were merely skeletons clothed in dirty rags of canvass." Borthwick stayed "in a holey old canvass hotel, which freely admitted both wind and water." He was amused that next to the hotel was a French doctor whose sign advertised that he was a Paris physician, dentist, pharmacist, and killer of rats.

Mokelumne Hill's population eventually topped 15,000 (today it is 650) and became one of the most ethnically diverse communities in the Mother Lode. There were Anglo Americans, African Americans, Germans, Italians, English, Spanish, French, Mexicans, Chileans, Chinese, and Irish. At its height the community boasted a wide variety of mercantile enterprises to service the swarming miners, including food markets, dry goods and clothing emporiums, hotels, saloons, gambling halls, livery stables, breweries, liquor stores, cigar shops, carpenters, and wheelwrights. And Mokelumne Hill provided numerous venues for amusement: improvised theaters, saloons, and gambling halls.

A young man, H. Q. Clark of the Adams Express Company, found his daily routine dreary and wished for the excitement of acting in a theater, any theater. He organized a theatrical company in Mokelumne Hill, rented a building, and constructed a stage. To light the stage, candles were arrayed as footlights on a board that could be raised and lowered to achieve the desired effect. For a production of Shakespeare's *Richard III* in 1853, Clark pestered the company into allowing him to star as Richard. Clark was particularly eager to perform the famous Bosworth Field scene, which features the legendary "A horse! a horse! my kingdom for a horse!" passage. As James Ayers remembered,

> The house was crowded. It was a dark stage. Richard was writhing on his couch. . . . Falling on his knees, he cried out to the people in front to bind up his wounds and give him another horse. As he made this appeal in tremulous tones

> a musical burro [a device which mimicked the braying of a mule] which one of the boys had mischievously fastened under the stage answered his prayer in corrugated notes that made the rafters shake. A great roar went up from the audience.

A stagehand who could not see the front of the stage then thought they had reached the prescribed moment for him to raise the footlights, and he cranked up the board, which promptly hit the erstwhile Richard III in the head, abruptly ending the performance and Clark's acting career.

By contrast, another well-liked entertainment, called "pedestrianism," was genteel, perfectly odd, and very popular. Participants were lionized, featured on collectible cards (akin to today's baseball trading cards), and made the subjects of hero-worshipping biographies.

In these pedestrian events, which often took place on racetracks that also served horse races and bull-and-bear fights, participants would attempt to walk a certain distance in a specified time, such as five hundred miles in five hundred consecutive hours. In the early 1850s, San Francisco had two such racetracks, and spectators numbered in the hundreds and sometimes thousands. Members of the audience would wager, and the "walker" would be paid upon successful completion of the task. Occasionally, the wager would require the pedestrian to perform additional assignments while walking, as was the case for a Mr. Grisgby, who in September 1854 was challenged to complete the following undertakings within one hour: walk a mile, pull a wagon two miles, walk backward one mile, and, finally, pick up fifty stones one yard apart and deposit them in a basket. Grigsby finished the task in fifty-nine minutes and thirty-six seconds and collected $500 for his "pedestrian feat," as the *San Francisco Commercial Advertiser* punned.

A more standard exhibition was undertaken by a Mr. Kelly of Sacramento. On June 6, 1852, the *San Francisco Daily Alta California* reported, "A man named Kelly commenced on Sunday morning, at Selby's ranch, below Sacramento, the arduous feat of walking a thousand miles in a thousand consecutive hours, on a wager of $2,000. He came in at the end of the tenth mile last evening as fresh as when he started on his long journey." Three weeks later, the *Daily Alta*

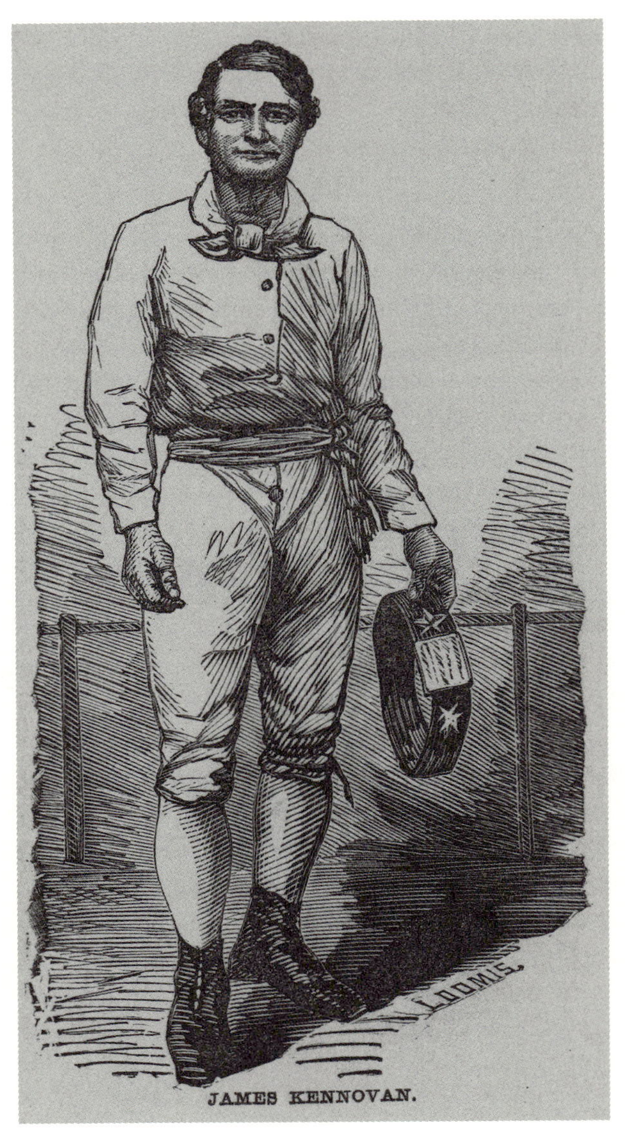

James "Uncle Jimmy" Kennovan, champion pedestrian, frequently concluded his competitive walks by dancing a jig. Illustration from *Fact and Fancy in the Life of a Pedestrian: Being a Full History of the Champion Pedestrian of the World, James Kennovan of San Francisco,* by J. A. Woodson (San Francisco: Sullivan, 1863). Courtesy of the California State Library, Sacramento; California History Section.

California provided an update on Kelly's progress: "Kelly has walked 710 miles of his 1000. His feet, which a few days since were much blistered, are now nearly well, and there is no reason to doubt that he will accomplish his arduous task." On July 15, the newspaper offered a one-sentence summary of the six-week trek: "Kelly, the pedestrian, has succeeded in accomplishing the great feat of walking 1,000 miles in 1,000 consecutive hours."

Foot races also fell under the umbrella of "pedestrianism." On August 13, 1851, hundreds flocked to San Francisco's Pioneer Race Course to observe a foot race between two men. They would race twenty-five yards, with the victor receiving $200 ($5,750 in today's money). To the manifest dissatisfaction of the crowd, the champion won the twenty-five-yard foot race over his sluggish opponent by *eight yards*.

But perhaps the final word on pedestrianism came three years later, on December 18, 1854, when the *Daily Alta California* conveyed this news from Mokelumne Hill:

> The "great Western" walker Wheeler has just completed his feat of walking one hundred consecutive hours in one of the mining towns, and we see by a paragraph in the Calaveras *Chronicle*, that Peeler, the "great Southern" walker, was up last week for a walk of one hundred and two hours. What is accomplished by these exhibitions of pedestrianism we do not see.

Other diversions were primal and brutal. The most notorious was the bull-and-bear fight. This battle to the death pitted an enraged long-horned bull against a captured grizzly bear, while paying customers observed, and often bet, on the gruesome outcome. These popular contests were held in a pen ringed by a sturdy wooden fence and circled by hastily constructed amphitheaters that could house dozens to thousands of spectators. Every major mining camp and town staged these battles. Mostly these gory conflicts occurred on Sundays for an admission price of between $2 and $10 ($60 and $300 today). These blood-spattered events were accompanied by parades, music, horse races, and cock fights, all preceding the main event showcasing the four-legged warriors.

John David Borthwick described the human kaleidoscope at a Mokelumne Hill bull-and-bear fight in 1853:

> The scene . . . was one which would have made a crowded opera-house appear gloomy and dull in comparison. The shelving bank of human beings which encircled the place was like a mass of bright flowers. The most conspicuous objects were the shirts of the miners, red, white, and blue being the fashionable colours, among which appeared bronzed and bearded faces under hats of every hue; revolvers and silver-handled bowie-knives glanced in the bright sunshine, and among the crowd were numbers of gay Mexican blankets, and red and blue French bonnets, while here and there the fair sex was represented by a few Mexican women in snowy-white dresses, puffing their cigaritas in delightful anticipation of the exciting scene which was to be enacted.

In a letter written on July 5, 1851, twenty-two-year-old Joseph Smith Hill, a forty-niner from Iowa, recalled a contest in Nevada City:

> They brought out . . . [a] bull and Grizly Bear and chained them together, the chain being long enough for them to get twelve feet apart. They got the Grizly very mad befor they let him out of his cage so that he would take after the bull, which he did as soon as he was turned out. But being a very small one, did not hurt the bull much, and soon he got tired out, and the bull would horn him & throw him against the pen, and make the old bear bawl worse than a calf. . . . They have another fight tomorrow.

Borthwick told of a much fiercer affair in Mokelumne Hill. The fight featured "General Scott, the Celebrated Bull-Killing Bear," which he described as "a grizzly bear of pretty large size, weighing about twelve hundred pounds." The Celebrated General Scott was unceremoniously dumped from his cage into the fighting ring, and then the skittish bull was led into the pen.

> [The bull] made up his mind to fight; and after looking steadily at the bear for a few minutes as if taking aim at him, he put down his head and charged furiously at him across the arena. The bear received him crouching down as low as he could, and though one could hear the bump of the bull's head and horns upon his ribs, he was quick enough to seize the bull by the nose before he could retreat. . . . The bear, lying on his back, held the bull's nose firmly between his teeth, and embraced him round the neck with his fore-paws, while the bull made the most of his opportunities in stamping on the bear with his hind-feet. At last the General became exasperated at such treatment, and shook the bull savagely by the nose.

The bull managed to break free as the crowd roundly cheered the bear's strength and the bull's resilience. The bloodied bull eyed the bear briefly and then made another rush at his opponent

> Again poor bruin's ribs resounded, but again he took the bull's nose into chancery, having seized him just as before. The bull, however, quickly disengaged himself, and was making off, when the General . . . seized his hind-foot between his teeth, and, holding on by his paws as well, was thus dragged round the ring before he quitted his hold. . . . It was thought that the bull might have a chance after all. He had been severely punished, however; his nose and lips were a mass of bloody shreds, and he lay down to recover himself.

A vote of the raucous onlookers agreed to introduce a second bull to the fight, but even two against one, the grizzly bear soon had the advantage over the tiring and mutilated bulls. Clearly, General Scott had won the day. The competitors were disentangled, the bear was led back to his cage, and the two bulls were shot. Borthwick observed, "A man sitting next me, who was a connoisseur in bear-fights, and passionately fond of the amusement, informed me that this was 'the finest fight ever fit in the country.'"

"SO WONDERFUL, SO DANGEROUS, SO MAGNIFICENT A CHAOS"

THE 1849 CALIFORNIA CONSTITUTION

In late August 1849, a remarkable caravan slowly rolled into Monterey, on California's central coast. Throughout the week, by ship, horseback, wagon, and on foot, forty-eight men descended on this commercial and social hub of the Gold Rush. The crude public lodgings of the small seaside district were inadequate for the dozens of new arrivals, many of whom carried their own bedrolls and camped under the stars. At noon on September 1, 1849, these forty-eight men got down to business, gathering in the only building in California that could accommodate their conclave: a two-story stone structure called Colton Hall. What they achieved over the next six weeks would transform California.

By mid-1849, Californians had endured months of lawlessness amid the unwillingness of Congress to appoint an effective provisional government. Frustrated, California leaders called for a solution—now. As the *Alta California* wrote on May 3, "The present state of anarchy (for we can call it nothing less) is much to be deplored and is easily remedied, by united, vigorous and immediate action." On June 3, California's military governor, Brigadier General Bennet Riley, issued a "Proclamation to the People of California" calling for the election of delegates to a constitutional convention, whose task would be to codify some form of government for the future California, whether as a United States territory or as a state. The forty-eight people elected were all men and mostly white, although seven were Californios—only two of whom were fluent in English—and the group included eleven delegates from the southern section of California in addition to the thirty-seven hailing from the northern mining camps and cities. There were men from states including New York, Missouri, Louisiana, Maryland, New Jersey, Virginia, Massachusetts, Connecticut, Florida, Illinois, Indiana, Ohio, Pennsylvania, Texas, and Wisconsin, as well as immigrants from France, Ireland, Scotland, Switzerland, and Spain. Of the American-born delegates, more were from slave states than free states. Thirty-three delegates had been in California for less than three years, thirteen had resided in the region for less than a year, and more than half were under the age of thirty-five. The oldest delegate, José Antonio Carrillo of San Francisco, was fifty-three. When the participants listed their occupations for the official record they identified as lawyers, farmers (although some preferred the term "agriculturist"), merchants, military officers, printers, physicians, and bankers. Carrillo chose the label "labrador," or farmer/farmworker. Pedro Sansevaine, originally from Bordeaux, identified himself as a "negotiant," and Benjamin Moore listed his profession as "elegant leisure." Also participating was perhaps the best-known figure in California, John Sutter, the domineering land baron of New Helvetia, who described himself as a "farmer." Interestingly, not a single one classified himself as a miner.

Upon convening, the assemblage strongly favored statehood for California; wishing to exercise as much local control over their affairs as possible, the vast majority of the delegates rejected applying for territorial status. A few cautious souls balked, pointing out that a region securing statehood without first passing through territorial status had only been done once previously—Texas in 1845—and advocates for a Territory of California argued that, in its formative stage, California might be better served with the federal government footing the bill for governance. But the statehood advocates were weary of congressional procrastination and easily emerged victorious as the members voted to join the union.

Their attention then turned to the most acrimonious concern in United States history: slavery. The issue bubbled with fury and discord

Colton Hall, in Monterey, was the site of the 1849 California Constitutional Convention. Courtesy of the California State Library, Sacramento; California History Section.

throughout mid-nineteenth-century America, but at the convention the issue of slavery was surprisingly and quickly dispatched. With a unanimous vote, one of the first constitutional provisions approved was the declaration that "neither slavery, nor involuntary servitude, unless the punishment for crimes, shall ever be tolerated in this State."

This particular debate hinged not on the morality of slavery but on the virtue of individual labor. As Monterey alcalde Walter Colton wrote in his diary the preceding year: "All here are diggers, and free white diggers won't dig with slaves. They know they must dig themselves; they have come out here for that purpose, and they won't degrade their calling by associating it with slave-labor: self-preservation is the first law of nature." John Milner, a slaveholding gold seeker from Alabama, reinforced this notion when he complained, "With twenty good

Negroes and the power of managing them as at home, I could make from ten to twenty thousand dollars per month. But here a fellow has to knock it out with his own fist, or not at all."

But the ease of passage of the antislavery provision did not mean that black people—slave or free—were welcomed by all in Gold Rush California. Delegate Morton M. McCarver, originally of Kentucky but lately of Oregon, where he had served as speaker of the provisional legislature, introduced an amendment to the provision. It read, "Nor shall the introduction of free negroes, under indentures or otherwise, be allowed." The amendment fostered considerable debate and was eventually withdrawn, but McCarver did not capitulate. A few days later he proposed an addition to the California Bill of Rights on the status of African Americans. It prompted heated argument for days. McCarver's addendum read:

> The Legislature, shall, at its first session, pass such laws as will effectually prohibit free persons of color from emigrating to and settling in this State, and to effectually prevent the owners of slaves from bringing them into this State for the purpose of setting them free.

McCarver passionately argued for exclusion:

> No population that could be brought within the limits of our territory could be more repugnant to the feelings of the people or injurious to the prosperity of the community than free negroes. They are idle in their habits, difficult to be governed by the laws, thriftless, and uneducated. It is a species of population that this country should be particularly guarded against.

It is generally believed that many of the delegates supported a permanent ban on free blacks in California, but, for fear of jeopardizing Congress's approval of the new state constitution, the convention rejected the article, 31 votes to 8.

The 1849 Constitutional Convention also weighed and passed a ban on state-run lotteries and the sale of lottery tickets. Delegate Roman

Price of San Francisco thought the lottery ban was absurd because, as he asserted, "The people of California are essentially a gambling people." Henry Halleck, a representative from Monterey, countered, "We may be a gambling community, but let us not in this constitution create a gambling state." The lottery ban was adopted, but private gambling went unmentioned in the document.

Delegates tried to tame some of the wilder impulses of Gold Rush Californians through the new constitution, and it was in this vein that they prohibited dueling. Although a time-honored means of restoring reputations in the United States, the practice was falling out of favor back East, and with many of the convention members recent arrivals from states where dueling was prohibited, they felt the ban should extend to California, where it was still practiced. The proviso stated that any citizen who should "fight a duel with deadly weapons, or send, or accept a challenge to fight a duel, . . . or knowingly assist in any manner those [who do]," should "not be allowed to hold any office of profit, or to enjoy the right of suffrage." Ironically, one of the strongest proponents for a dueling proscription was William Gwin of San Francisco (later a United States senator from California), who participated in a duel with United States representative J. W. McCorkle in 1853. Angered over purported insults, Gwin and McCorkle took thirty paces, wheeled, and then shot at each other with rifles. They both missed. As *Frank Leslie's Popular Monthly* magazine recalled in 1877, after "three ineffectual shots" were discharged, a bystander rushed forward to inform the duelists that they had received false information and that the harsh words that had precipitated the duel had all been a misunderstanding. There were apologies. No arrests, no harm done.

Perhaps the most forward-looking provision debated during the 1849 California Convention concerned women's rights—specifically the property rights of married women. Throughout the United States, married women had few, if any, political and economic rights; custom and law favored supremacy of the husband in all such affairs. The California constitution, however, provided that the property of the wife was not to be merged with that of her husband but was to remain her own separate property. The convention had turned tradition on its head, and for a curious reason. Delegate Henry Halleck, a bachelor from Monterey, explained the thinking:

> I shall advocate this section in the Constitution, and I would call upon all the bachelors in this Convention to vote for it. I do not think we can offer a greater inducement for women of fortune to come to California. It is the very best provision to get us wives that we can introduce into the Constitution.

This section led some delegates to bemoan what they interpreted as a gratuitous attack on traditional marital relationships. Charles Botts, a representative from Monterey, urged the convention to reverse course and "expunge [the clause] altogether from the Constitution . . . because I think it is radically wrong." He maintained that

> there is no provision so beautiful in the common law, so admirable and beneficial, as that which regulates this sacred contract between man and wife. Sir, the God of nature made woman frail, lovely, and dependant; and such the common law pronounces her. . . . I say, sir, the husband will take better care of the wife, provide for her better and protect her better than the law. He who would not let the winds of heaven too rudely touch her, [he] is her best protector. When she trust him with her happiness, she may well trust him with her gold. . . . This doctrine of woman's rights is the doctrine of . . . mental hermaphrodites.

Regardless of the reasoning or the peculiar matrimonial motivations, this constitutional provision was remarkably freethinking for the era, and, as historian William Henry Ellison concluded when reconsidering the event on its one-hundredth anniversary, it is "evidence of the vision and liberalism of the constitution makers of California."

The convention drew to a close on October 12, six weeks after it had begun. The work concluded, it was time to celebrate, and each delegate contributed $25 for food, decorations, and entertainment. Colton Hall, where the men had been meeting, was adorned with pine boughs. Three improvised, ersatz chandeliers lit the dance floor and, at eight o'clock in the evening, the guests began to congregate. In addition to the convention delegates, there were other men from Monterey, and also present were those rarest jewels of the Gold Rush—women. Newspaper correspondent Bayard Taylor described the scene:

> There were sixty or seventy ladies present, and an equal number of gentlemen, in addition to the members of the Convention. The dark-eyed daughters of Monterey, Los Angeles and Santa Barbara mingled in pleasing contrast with the fairer bloom of the trans-Nevadian belles. The variety of feature and complexion was fully equalled by the variety of dress. In the whirl of the waltz, a plain, dark, nun-like robe would be followed by one pink satin and gauze; next, perhaps, a bodice of scarlet velvet with gold buttons, and then a rich figured brocade, such as one sees on the stately dames of Titian.

The celebrants danced to a band that, Taylor noted, "consisted of two violins and two guitars, whose music made up in spirit what it lacked in skill." At midnight, dinner was served:

> The refreshments consisted of turkey, roast pig, beef, tongue and *patés*, with wines and liquors of various sorts, and coffee. A large supply had been provided, but after everybody was served, there was not much remaining. The ladies began to leave about two o'clock, but when I came away, an hour later, the dance was still going on with spirit.

The next morning, the delegates endorsed the constitution. When all had signed, a cannon barrage thundered at the Monterey fort and there was a thirty-one gun salute: thirty cannon blasts for each state that joined the union before California, and a final one for the soon-to-be thirty-first star on the flag. John Sutter gleefully exclaimed, "Gentlemen, this is the happiest day of my life. It makes me glad to hear those cannon. . . . This is a great day for California!"

The leading nineteenth-century chronicler of California history, Hubert Howe Bancroft, later noted, "Never in the history of the world did a similar convention come together. They were there to form a state out of unorganized territory; out of territory only lately wrested from a subjugated people, who were allowed to assist in framing a constitution in conformity with the political views of the conquerors." Bayard Taylor, then a twenty-four-year-old journalist reporting for Horace Greeley's *New York Tribune,* also marveled at their accomplishment: "The members of the Convention may have made some blunders in the course of their deliberations . . . but was there ever a body convened under such peculiar circumstances? Was there ever such harmony evolved out of so wonderful, so dangerous, so magnificent a chaos?"

On November 13, 1849, California voted to ratify its new constitution and to form the state's first government. For those in influential positions in the society, such as landowners, merchants, and the assembled delegates, the constitution mattered quite a bit, but in general most residents were either unaware that a constitution had been crafted and/or did not know an election was being held and/or perhaps, just perhaps, they did not care. Out of 110,000 eligible voters, only 12,061 votes were cast. To many of the transient gold seekers, a new constitution was of no consequence.

THE LEGISLATURE OF A THOUSAND DRINKS

THOMAS JEFFERSON GREEN AND THE FIRST CALIFORNIA LEGISLATURE

In 1835, West Point graduate Thomas Jefferson Green moved to the nascent republic of Texas with the intent of founding a town in the wide-open landscape. More than new towns, however, what the republic needed was military leadership, and Texas officials offered Green a commission in the Texas Army.

In 1842, while on an expedition along the contentious Mexican border, Green and forty Texas soldiers were captured by Mexican forces. He and his compatriots were assigned to a prison near Mexico City notorious not only for prisoner escapes but for the escapees being captured and then executed. After nearly three years of confinement, Green himself managed to escape, and he returned unscathed to Texas just as the territory was being annexed as a state.

When the Gold Rush began, Green contracted gold fever and headed to Sacramento. He had been in the river city only a few months before he was elected to the new California Senate; perhaps his new neighbors were impressed by his lengthy legislative resumé: Green had already served in the legislative bodies of North Carolina, the Territory of Florida, and the Republic of Texas.

California did not officially become a state until September 9, 1850, but the California Constitutional Convention of 1849 decided that they could establish a self-governing dominion until formal ratification. Although the legislators represented some of the diversity of California's demographics, they also reflected popular biases. There were no women, Native Californians, African Americans, or immigrants from non-European nations. A handful were established residents in the form of Californios and holders of Mexican land grants, but many more were recent arrivals in California from states including New York, Missouri, Alabama, Kentucky, Texas, Ohio, Indiana, Michigan, Virginia, Connecticut, Tennessee, and South Carolina. There were farmers, merchants, and a smattering of lawyers. Green was among sixteen senators and thirty-six assemblymen elected to the legislature and scheduled to take their oaths of office in El Pueblo de San José de Guadalupe, the new state capital, on December 15, 1849.

But in December of 1849, San José was a suburb of Hell. The Santa Clara Valley town of four thousand was in the midst of a six-month season of torrential rain, a series of deluges that would drop thirty-six inches of precipitation. The roads were quagmires and virtually impassable. The best accommodation in town was the City Hotel, an overcrowded wood-frame building, one and a half stories tall. There were only enough rooms to accommodate one-quarter of the boarders. At night, space on the dining room and barroom floors was available for $5 in gold. As historian Frederic Hall wrote, the lodgers also discovered that they had company, "joint-tenants in the house" who were not registered and could not "be ejected by the law, except the law of self-preservation": legions of fleas. "If a man scratched his head," Hall continued, "nobody for a moment supposed it was for an idea."

Drinking water had an odd, salty taste, although most did not care as there were plenty of other libations available. San José was a city more receptive to John Barleycorn than God, as saloons far outnumbered churches. Prostitutes were readily available, and, as local legend declared, when more were needed, wagons were dispatched to San Francisco to fetch a new supply. When dinner was served, historian

Hall noted, there was a mad scramble: "It was a hazardous undertaking. . . . The rush was so great, that crowding through the dining-room door put one in mind of trying to drive a four-horse team through a single door of a stable."

The members of California's new government straggled in. With terrible weather, treacherous roads, and horseback being the only reliable means of transportation, many of the legislators had difficulty reaching San José at all, and the first two sessions of the first legislature adjourned for lack of a quorum. When the Constitutional Convention selected San José for the capital, San José residents had promised the delegates that a suitable building for the legislature would be provided, but by the time the event was upon them, San José could not deliver as promised. Arriving delegates were greeted by a barely useable unfinished adobe box. Historian Hubert Howe Bancroft later described the capitol this way:

> Sixty feet long and forty feet wide, [it was] two stories in height, having a piazza in front. The upper story, devoted to the use of the assembly, was simply one large room, approached by a flight of stairs from the senate-chamber, a hall forty by twenty feet on the ground-floor; the remainder of the space being occupied by the rooms of the secretary of state, and various committees. For the first few weeks, owing to the incompleteness of their hall, the senators held their meetings in the house of Isaac Branham, on the southwest corner of the plaza.

The chambers were unanimously deemed unsatisfactory, and one of the very first acts passed by the legislature was to immediately move the capital to Monterey and start again. However, due to the very muddy roads and the inability to move anything anywhere, the legislature tabled the measure and proceeded with business.

As chairman of the Senate Finance Committee, Thomas Jefferson Green was tasked with finding some source of revenue for the new state. He shepherded legislation creating bonds, property taxes, a poll tax, and the initially popular but later wildly *un*popular Foreign Miners' Tax of 1850. Green also authored legislation to establish the

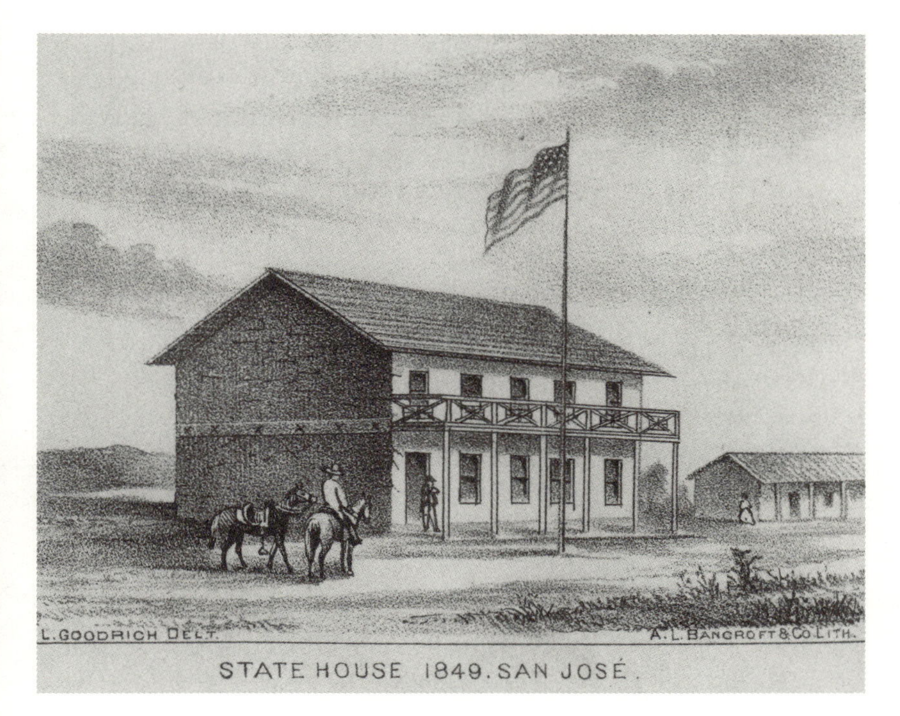

California's first state house, San José, 1849. San José was the capital of California until 1852. Drawing from *The History of San José and Surroundings*, by Frederic Hall (San Francisco: A. L. Bancroft and Co., 1871). Courtesy of the California State Library, Sacramento; California History Section.

College (later University) of California and was an early benefactor of the California State Library.

But Green's tenure is best remembered for the nickname he applied to the first California legislature. He called it "The Legislature of a Thousand Drinks." With his lengthy legislative experience, Green appreciated the value of alcohol-fueled lobbying, and he encouraged his colleagues to meet regularly in the local taverns. When a session concluded, Green would loudly declare, "Let's have a drink! Let's have a thousand drinks!" It did not take much effort to persuade the lawmakers to comply.

Green's suggestion for his colleagues to drink up was well taken, at least as far as it applied to government business. In 1885, California scholar Theodore Hittell concluded that "it is certain that no legislature has ever sat in the state that did more work, more important work, or better work." In 1888, historian Hubert Howe Bancroft shared a similar but more colorfully expressed opinion:

> It has always been alleged that the American-Californians of an early period drank freely, and this body has been styled the 'legislature of a thousand drinks.' However this may have been, it was the best legislature California ever had. For what they drank, the members returned thanks. All were honest—there was nothing to steal. Their pay was no inducement, as they could make thrice as much elsewhere. Furthermore, this was before Californians began to sell themselves as political prostitutes.

In 1988, a plaza was constructed in downtown San Jose to commemorate the site of the state's first capital. Carved in granite, next to a giant brass Seal of the State of California, are the words "The Legislature of a Thousand Drinks."

"AN ACT FOR THE BETTER REGULATION OF THE MINES AND THE GOVERNMENT OF FOREIGN MINERS"

THE FOREIGN MINERS' TAX AND THE FRENCH REVOLUTION

Something had to be done immediately or there would be hell to pay, warned the California State Senate's Finance Committee chairman on March 15, 1850. Senator Thomas Jefferson Green of Sacramento fervently exclaimed that the Gold Rush had "excited the wildest cupidity which threatens California with an emigration overwhelming in number and dangerous in character." Thousands of "foreigners"—that is, those not born in the United States—were descending daily, and they represented "the worst population of the Mexican and South American States, New South Wales, and the Southern Islands to say nothing of the vast numbers from Europe." Worse yet, noted Green, "the convicts of Mexico, Chili, and Botany Bay are daily turned upon our shores who seek and possess themselves of the best places for gold digging." Legislation was necessary, now, Green argued, to regulate the mines and the alleged pernicious influence of this horde of foreign miners. Not only was the presence of these miners detrimental to the economy, the senator argued, but it was also harmful to the community, as the "low state of morality which such a population spreads broadcast in the land is to be deeply lamented."

Senator Green and the Finance Committee urged the adoption of "An Act for the Better Regulation of the Mines and the Government of Foreign Miners," then under consideration in the legislature. This proposed law would place a monthly tax on foreign miners as a mechanism to control this burgeoning population. "The bill," said Green, "will do much to remedy these apprehended evils," as the act "requires the foreigner upon the plainest principles of justice to pay a small bonus for the privilege of taking from our country the vast treasure to which they have no right."

The few critics of the proposed tax noted that while the professed purpose was to raise revenue, the law was a blatant xenophobic weapon designed to purge the goldfields of foreign-born miners through economic harassment. The Foreign Miners' Tax was fiscally illogical as well, opponents argued. With foreigners comprising about one-quarter of the population, if they refused to pay the tax or left the mining camps, the much-needed revenue would actually plummet. Merchants would be harmed, as foreigners would have less money available to purchase provisions. Despite these objections, however, the bill easily passed a few weeks later.

Thomas Jefferson Green predicted that those affected would "cheerfully" pay the tax. He was wrong. There was resentment, anger, and violent reaction. And the act spawned the peculiar incident known as the "French Revolution."

Thousands of people from around the world were drawn by the lure of gold, and as a group they imbued the goldfields with a diverse and cosmopolitan air. The French showed particular enthusiasm due to the economic depression and political uncertainty back home, and between 1849 and 1852 an estimated thirty thousand French citizens migrated to Gold Country, where they came to comprise roughly 10 percent of the state's population. National rivalries and cultural differences led to tension and hostility between various groups, and the French were not exempt.

When the Foreign Miners' Tax was put into effect, it was immediately denounced by those the law initially targeted, namely the Mexicans, the Chileans, and the French. Many did not object to the tax *per se* but felt that the $20 monthly fee was exorbitant and that a lower monthly assessment, perhaps $4 or $5, would be more fair. Benjamin Butler Harris, a young lawyer originally from Tennessee and then a resident of Sonora, Tuolumne County, characterized the tax law as "prohibitive and so outrageously extortionist in its terms as to make an angel weep." As predicted, the tax did harm businesses in regions that had the greatest proportion of foreign miners. William Perkins, a merchant in Sonora, called the tax a "foolish act . . . [that] is operating badly against the commercial interests."

Petitions calling for the act's repeal or modification were circulated, and "indignation meetings" commenced when tax collector Lorenzo Besançon arrived in Tuolumne County in May 1850 to assume his appointed duties. Defiant assemblies of Mexican, Chilean, French, and a sprinkling of English and German miners gathered in American Camp (soon be rechristened Columbia) and in Murphy's New Diggings (later called Murphys) to await the arrival of Besançon and his deputies. All foreign miners were spitting mad, and the French in particular angrily pledged that they would not pay one cent of the tax.

In his 1856 book *Californische Skizzen*, Friedrich Gerstäcker, a native of Hamburg, Germany, set the scene:

> From different sections of the town . . . discordant and strange sounds reached the ear; the boom of a Chinese gong mingled with the sharp notes of a child's trumpet. . . . In this improvised and raucous concert the donkeys of the miners joined, to which must be added the chopping of wood and the refrains of French songs coming from nearby tents or brush. Groups came and went . . . from tent to tent, laughing and gesticulating; twilight settled down; a few light clouds, gilded by the setting sun, floated in the sky; it was a scene to make even such poor devils forget the fatigues of the day and their worries for the morrow.

The scene he described in his book, later translated and published as *Scenes of Life in California*, was just one of many similar gatherings protesting the coming of the taxman. Demonstrations punctuated with rebellious unfurlings of Chilean, Mexican, and French flags roiled the region. A few Americans sympathized with the foreign miners, but most were furious at the dissidents threatening insurrection.

It all came to a head on Sunday, May 19, 1850. Delegates from the various assemblies descended upon Sonora to meet with the taxing authorities, specifically tax collector Lorenzo Besançon. The deputations sought a clarification of the law and hoped to determine if amendments were feasible. Various reports observed that these emissaries were accompanied by several thousand French, Chilean, and Mexican supporters. Most proceeded peacefully, but at least a handful, flush with resentment, threatened to burn down the town. Benjamin Butler Harris remarked that "hot, boiling speeches, resolutions and threats were made" and that Frenchman Casimir Lebetoure, a Sonora merchant, a member of the town council, and William Perkins's next-door neighbor, was the ringleader. Lebetoure made an incendiary "liberty or death" oration in French and Spanish to the eager throng and dismissed the American miners as merely "servile peons sent here by their masters to delve in the mines."

There was a potential flashpoint when an American attending the parley with Collector Besançon elbowed his way through the crowd while departing the gathering. Angered by such rudeness, a Chilean protestor drew a pistol on him. William Perkins noted that, in his estimation, "the whole foreign population of the town assumed a threatening attitude." Sonora residents armed themselves, just in case, and sent messengers to nearby communities requesting assistance. Five hundred riflemen arrived within five hours and, as Perkins recalled, "marched into the town like disciplined soldiers." The protesting assembly withdrew and the state of affairs cooled. Briefly.

The next day, May 20, 1850, word reached the mustered militia in Sonora that at a foreign miners' encampment in Columbia, some seven miles from town, the residents had torn down the American flag and replaced it with the French banner. As Perkins recounted, the armed Sonora troop then "started to the camp in question to chastise the insult." Upon arrival the angry band found "every thing quiet and [everyone] had the good sense to refrain from any wanton aggression."

Benjamin Butler Harris recalled something quite different indeed.

"The convention," he wrote, "was transferred to Columbia, when, at an early hour, the Mexican and French flags were raised high." When the "six hundred American men came in view from towards Sonora," they were "armed with guns, some with axes, picks, clubs, shovels, and some only with naked fists." Upon arrival at the camp, "the foreigners, with streamers, banners, and flags were parading the streets of Columbia, preceded by a naked Indian daubed in war paint, all singing the *Marseillaise* and other revolutionary songs." The Americans rushed at the encamped protestors, sending the foreigners scattering in all directions. Harris boasted that the six hundred American attackers forced nearly four thousand French and Mexican men to scramble into the hills, and they intimidated some protestors into feigning they were asleep, some even snoring loudly as a way of proving they were not involved and should therefore not be detained by the marauding militia. The Americans, wrote Harris, quickly disarmed their opposition and proceeded to pepper the French and Mexican flags with many volleys from their rifles. The swaggering throng removed the offending banners and "hoisted and cheered the Stars and Stripes."

Meanwhile, in Murphys, a herald fresh from Sonora announced that two Frenchmen and a German had been arrested in the uprising and were awaiting lynching, and that the sheriff had been murdered by a Spaniard. The camp, Friedrich Gerstäcker recalled, was instantly animated by a "warlike enthusiasm." As a council of war gathered, a German named Fuchs declared the reported arrests "a horrible outrage that cries for vengeance—that can only be washed out in blood." Gerstäcker recounted that the community resolved to march to Sonora to rescue their brethren, but that

> great confusion prevailed at the moment; the greater part of the combatants, to give themselves courage without doubt, had consumed more copious libations of wine and brandy than usual. Germans, Spaniards, French were in the crowd, each talking his own language, without too clearly understanding his neighbor, for from time to time some phrases in English, singularly mutilated, became necessary to serve as a means of interpretation.

While a few called for "cool-headedness" and verification of the rumors before advancing, most prepared for the offensive. A young French woman demonstrated a noteworthy zeal for combat. Known to history as Madame Louis, she was, wrote Gerstäcker, a dealer in a gambling hall and "a rather feminine figure of about twenty-six years, with black eyes, tall and slim." Madame Louis had outfitted for battle in a "complete Amazon costume; she wore dark colored pantaloons, a red woolen shirt, a large felt sombrero over one ear, decorated with an ostrich plume; she had a double-barreled shotgun slung over her shoulder, two pistols and a hunting-knife in her sash." She harangued the reluctant and the inebriated and called for action, action now! Madame Louis cried, "Let all who feel a courageous heart beating in their breast fall in under our flag. *Let us go, countrymen, the day of glory is here.*" She was joined by another woman, who had appeared suddenly and dramatically upon a horse, "her head covered by a felt hat, set off with a large red ribbon; and a silvery voice, strong with resolution." The expedition mobilized in their "original and picturesque aspect" of "more or less forbidding appearance, fairly uniformly clothed in red shirts and caps of the same color, or else hats of black or grey felt." As for arming themselves,

> everyone had seized the first weapon that fell to his hand; the greater part had double-barreled shotguns, some carbines, or simply muskets; many carried swords, daggers or pistols as side arms. A good many of these weapons were not in condition; regardless of rust covering them, they had been dragged forth for the occasion from the most obscure or dusty corners.

Now armed and motivated, the column, five hundred strong, headed toward Sonora, led by, as Friedrich Gerstäcker noted, "our two Amazons galloping, laughing, and singing as they went."

Upon arriving at the outskirts of Sonora, the procession discovered that the story of the Frenchmen "thrown into prison and in danger of death" was "a pure invention. . . . It was fully proved that all the excitement was the result of a monstrous canard conceived in

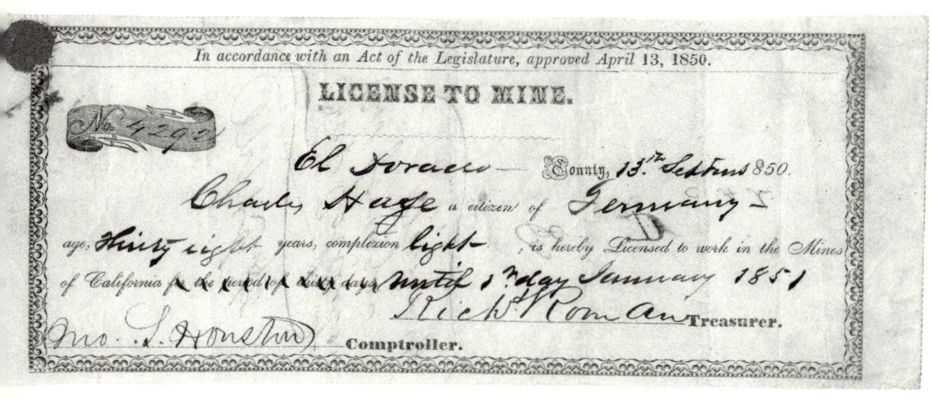
In accordance with an Act of the Legislature, approved April 13, 1850.

LICENSE TO MINE.

No. 4292

El Dorado County, 13th Septbr 1850.

Charles Hage a citizen of Germany

age, thirty eight years, complexion light, is hereby Licensed to work in the Mines of California for the period of thirty days until 1st day January 1851

Rich'd Roman Treasurer.

Jno. S. Houston Comptroller.

From 1850 to 1851 and again from 1852 to 1939, California levied a tax on foreign miners, requiring primarily non-English-speaking migrants to pay for licenses to mine in the state. The controversial tax sometimes led to violence. Courtesy of the California State Library, Sacramento; California History Section.

malevolence and spread with culpable imprudence. . . . All this adventure was based on nothing but a falsehood told as a joke." As Gerstäcker succinctly concluded, "The matter rested there and all this escapade received thereafter [was] the name of 'The French Revolution.'"

Embarrassed, the Murphys party retreated and, in fact, many of the foreign miners left Gold Country completely in the weeks to come. However, the ethnic and cultural antagonisms detonated by the tax confrontations did not dissipate, and for months afterward there were several violent incidents inspired by the law. In July 1850, a dozen men were murdered in the Sonora vicinity within a week, and throughout the remainder of 1850 and on into 1851, thefts and bloody confrontations surged in the Southern mines, where most of the foreign miners were based. In his 1851 book *Sixteen Months at the Gold Diggings*, Daniel Woods of Philadelphia attributed the carnage to the "heavy tax imposed upon the foreigners, which deprives many of them of employment. In consequence, they become desperate, often being destitute of the means with which to purchase their daily supplies. They are accordingly driven to steal and to murder."

In March 1851, mostly as a consequence of the increasing violence, the Foreign Miners' Tax of 1850 was repealed. But the idea did not expire with the law. In 1852, the California legislature revived it, but it also lowered the monthly assessment from $20 to only $3. In both 1853 and 1856, the act was amended again, to raise the tax levy to $4 and then $6 per month. By this time, the miners primarily targeted were the Chinese. Subsequently, the legislation was further modified to apply to all foreign workers, regardless of occupation, and between 1850 and 1870, the various configurations of the levy generated just over $5 million in revenue. Of that total, $4.9 million, or 98 percent, was paid by Chinese immigrants. The Foreign Miners' Tax of 1852 was, finally, officially repealed in April 1939, eighty-seven years after its adoption.

"WE ARE NOT THE DEGRADED RACE YOU WOULD MAKE US"

HAB WA, LONG ACHICK, NOMAN ASING, AND GOVERNOR JOHN BIGLER

The Chinese called the province Gold Mountain. It was a fabled land across the ocean, filled with gold, flush with opportunity, and offering unforgettable adventure. It was both destination and dream. Gold Mountain was California.

A few Chinese immigrants had arrived in California before the forty-niners, but during the Gold Rush their numbers increased dramatically. Census figures predating 1850, when California was granted statehood, are unreliable at best, but the numbers indicate that from 1840 to 1850 approximately 400 Chinese citizens immigrated to California. From 1850 to 1855, the number surpassed 27,000.

The vast majority of Chinese immigrants were male. They referred to themselves as "Gam Saan Haak"—Travelers to Gold Mountain—and considered themselves not settlers but temporary visitors hoping to get rich quick in the goldfields and then return to their families in China. It is nearly impossible to ascertain how many actually returned to China, but evidence indicates the number was small. Most Chinese immigrants could not afford the passage home. As historian Judy

Yung memorably stated in her 1995 book *Unbound Feet*, husbands and wives were separated "anywhere from ten years to a lifetime." A Cantonese folk rhyme cautioned:

> If you have a daughter, don't marry her to a Gold Mountain man.
> Out of ten years, he will not be in bed for one.
> The spider will spin webs on top of the bedposts,
> While dust fully covers one side of the bed.

Women of any race were rare in the Sierra Nevada and goldfields during the period, but Chinese women were particularly scarce. Of those 27,000 Chinese immigrants who arrived between 1850 and 1855, only 675 were women, and the difference became even more pronounced in the following decade. In 1862, Chinese immigrants totaled 7,214; only one was female.

Most of the newcomers gravitated to the mining areas, where the proportion of the population from China was significant. According to the census of 1860, 18 percent of the overall population of Sierra Nevada counties, and 35 percent of the foreign-born population, was Chinese. In some specific areas, the county concentration was even higher: Placer County was 22 percent Chinese, Amador County was 23.5 percent, and Mariposa County was 30 percent. Records from the 1860 census indicate that 80 to 90 percent of Chinese immigrants were engaged in mining, mostly placer mining. In the Southern Mother Lode, most Chinese newcomers passed through one village before dispersing elsewhere, and that Tuolumne County town soon became known as Chinese Camp.

Despite their numbers, or perhaps because of them, the Chinese were regarded warily by the Anglo culture and were considered unassimilable. People from other cultures were viewed suspiciously as well, but the Chinese were considered especially "alien" in their habits, language, and dress, and even in their time-tested mining technique of reworking abandoned claims. As scientist and explorer William Brewer noted in 1860, "They come here a 'peculiar people,' and stay so: they very seldom learn the language, and they adopt none of the customs of the country. They come with all the faults and vices of a heathen people, and these are retained here. . . . The morals of this class are anything but pure. All the vices of heathendom are practiced." Due to this cultural misconception, Chinese immigrants were also considered excessively clannish and secretive, clustering together for both support and protection. Their settlements had merchants, newspapers, barbers, doctors, herbalists, and assayers, all catering almost exclusively to the Chinese residents, and there were Chinese boarding houses, temples, theaters, cemeteries, and storefront gathering places that served as communication hubs and social centers. Through it all, the Chinese preserved their cultural identity while adjusting to a strange new land. This exclusiveness was reviled by Anglo miners, who frequently perceived the Chinatowns as threats to public safety and health. In fact, this banding was largely precipitated by the Anglo distrust and misunderstanding of Chinese culture in the first place.

The Chinese population faced restriction and regulation from the earliest days of the rush, and the oppression didn't let up in the years that followed. In 1852, the California legislature passed a bill that was designed to discourage Chinese immigration through the use of a "commutation tax" that required ship captains to submit a list of all foreign passengers and post a $500 bond for each of them. The bond could be waived, or "commuted," by paying a fee ranging from $5 to $500, collected from the passengers themselves. A few years later, in 1855, an act levied a $50 tax on every immigrant not eligible for citizenship, which included the Chinese. While these statutes were carefully phrased to apply to all foreigners, the context of the debate indicates that Chinese immigrants were the main target.

Various incarnations of a foreign miners' tax also focused on Chinese people, beginning in April 1850. This state law required foreign miners to pay $20 monthly for the privilege of searching for gold. County treasuries were the primary beneficiaries of the tax revenues. Statistics show that in the act's earliest form between 1850 and 1854, Chinese workers paid 50 percent of the tax collected, with the remainder spread mostly among Mexicans, Chileans, Peruvians, and the French. The original Foreign Miners' Tax was repealed eleven months after its

"The 'Heathen Chinee' Prospecting, Calif., Year 1852," photograph by Eadweard Muybridge, c. 1871. An early innovator of photographic technology, Muybridge often captioned his photographs with commonly used stereotypical ethnic descriptions. Courtesy of the California State Library, Sacramento; California History Section.

passage, only to be replaced by a later law that, although it reduced the monthly fee, was clearly meant to harm the Chinese, who were increasingly viewed as the most threatening and unwelcome of the immigrants. Records indicate that from 1854 until the Foreign Miners' Tax was declared unconstitutional in 1870, the Chinese paid 98 percent of the tax, a total approaching $5 million.

Worse than the discriminatory taxation was the violence Chinese miners were subjected to. Anglo miners attempted to exclude Chinese immigrants from the diggings throughout the goldfields, both in the Northern and Southern Mines. In the spring of 1852, as part and parcel of widespread anti-Chinese hostility during that year, Chinese mining camps along the American River were mercilessly raided. At Mormon Bar, sixty white miners accompanied by a brass band attacked two hundred Chinese residents. The marauders then rode down the river and assaulted four hundred Chinese miners working Horseshoe Bar. Despite the adverse intentions of the dominant culture, the Chinese continued to flood parts of the Mother Lode.

And when violence and taxation failed to stem the tide of Chinese immigration, the California power structure looked for a legal solution to their perceived problem. In 1852, California's third governor, John Bigler, responded to growing nativist sentiment by calling for restrictions on Chinese immigration. Bigler argued that controls were necessary, as, in his view, the Chinese were unable to assimilate as well as European immigrants. Additionally, Bigler wielded a very broad brush in his proposal and referred to all Chinese immigrants as "coolies," or low-status contract laborers akin to slaves. This term, first applied in the early 1700s to forced labor in India, was not only offensive, but it also did not accurately describe the makeup of the population of Chinese immigrants in California.

Bigler's comments led to the publication of two of the most remarkable documents of the Gold Rush: two letters written by respected Chinese leaders in response to Bigler and printed in leading periodicals of the time—the *Living Age* and the *Daily Alta California*, both headquartered in San Francisco.

The first letter, published on April 29, was composed by two Chinese merchants who were also officers in local benevolent societies, or "tongs." The authors were Hab Wa of the Sam Wo Company and Long Achick of the Ton Wo Company. They thoroughly objected to Bigler's use of the term "coolie" to describe them and their countrymen. "None are 'Coolies,'" they wrote, "if by that phrase you mean bound men or contract slaves." Hab Wa and Long Achick urged Governor Bigler to reconsider his characterization: "It has grieved us that you should publish so bad a character of us, and we wish that you could change your opinion and speak well of us in public." After all, they argued,

> there are no Chinese drunkards in your streets, nor convicts in your prisons, madmen in your hospitals, or others who are a charge to your state. They live orderly, work hard, and take care of themselves. . . . We are good men; we honor our parents; we take care of our children; we are industrious and peaceable; we trade much; we are trusted for small and large sums; we pay our debts; and, of course, must tell the truth.

In their final appeal to Governor Bigler they wrote, "We will only beg your Excellency not to be too hasty with us, to find us out and know us well, and then we are certain that you will not command your Legislature to make laws driving us out of your country. Let us stay here—the Americans are doing good to us, and we will do good to them."

Two weeks later, on May 5, 1852, a similar letter was published, this one written by a local Chinese leader and restaurant owner known as Noman Asing, whose Chinese name is believed to have been Yuen Sheng. The letter to "His Excellency Governor Bigler" references what Asing considered the relevant principles of the Declaration of Independence and the United States Constitution. He appealed to the governor's sense of history and fairness, but Asing also immediately took Bigler to task for his "official position," saying,

> the effect of your late message has been thus far to prejudice the public mind against my people, to enable those who wait the opportunity to hunt them down, and rob them of the rewards of their toil. You may not have meant that this should be the case, but you can see what will be the result of your propositions.

Asing also assailed Bigler's premise that excluding new Chinese residents would enhance the state's wealth:

> I have always considered that population was wealth; particularly a population of producers, of men who by the labor of

> their hands or intellect, enrich the warehouses or the granaries of the country with the products of nature and art. You are deeply convinced you say "that to enhance the prosperity and preserve the tranquility of this State, Asiatic immigration must be checked." This, your Excellency, is but one step towards a retrograde movement of the government, which, on reflection, you will discover; and which the citizens of this country ought never to tolerate.

Asing counseled Bigler to reevaluate his exclusion policy by considering the quality and scope of Chinese accomplishments through the centuries:

> We would beg to remind you that when your nation was a wilderness, and the nation from which you sprung *barbarous*, we exercised most of the arts and virtues of civilized life; that we are possessed of a language and a literature, and that men skilled in science and the arts are numerous among us; that the productions of our manufactories, our sail, and workshops, form no small share of the commerce of the world; and that for centuries, colleges, schools, charitable institutions, asylums, and hospitals, have been as common as in your own land.

Noman Asing concluded with this forceful statement:

> We are not the degraded race you would make us. We came amongst you as mechanics or traders, and following every honorable business of life. You do not find us pursuing occupations of degrading character. . . . You find us peculiarly peaceable and orderly. It does not cost your state much for our criminal prosecution. We apply less to your courts for redress, and so far as I know, there are none who are a charge upon the state, as paupers.

Regrettable to our modern sensibilities, Asing reinforced his case for the virtue of his countrymen by speaking favorably of the Chinese immigrants but to the detriment of other ethnic groups in California.

He denigrates the African American and Native American populations with commonly expressed racist stereotypes of the era, and he concludes that Chinese people would provide a more desirable population than members of those other groups.

The appeals of Hab Wa, Long Achick, and Noman Asing were disregarded, however. Within the next few years, the Foreign Miners' Tax was modified and increased. Chinese people were not allowed to testify in court. Violence against Chinese miners and communities continued unabated. Assaults became so widespread that a white correspondent to the *Placerville American* in 1857 asserted, "There ought to be a protection against the [Chinese] having to pay the onerous foreign miners' tax over three of four times; against sham licenses being given out and taken away from him, and his money extorted; and against being gagged, whipped and robbed whenever a worthless white rowdy chooses to abuse him thus, for pleasure or profit."

By 1862, the ongoing and often shocking abuse of California's Chinese residents led the state legislature to appoint a committee to investigate. It was designated the "Joint Select Committee Relative to the Chinese Population of the State of California." The committee chronicled the positive contributions of Chinese labor to the economy, but most famously it detailed the mistreatment of Chinese immigrants. Its final report contained a sobering list, provided by Chinese merchant associations, that catalogued eighty-eight murders of Chinese immigrants by white people within the previous few years, including eleven killings by Foreign Miners' Tax collectors. The report noted that only two of the suspects in the eighty-eight murders had been brought to trial. The committee's conclusion?

> It is a well known fact that there has been a wholesale system of wrong and outrage practiced upon the Chinese population of this state, which would disgrace the most barbarous nation upon earth.

"THE THEATRE OF SCENES WHICH WE HOPE WILL NEVER BE REPEATED" VIGILANTES

Men who arrived in Gold Country full of hope and enthusiasm often lost their initial dewy blush after a few weeks of back-breaking labor and no fortune to show for it. Disillusioned and destitute, some turned to crime. Chroniclers describe street brawls, robberies, bloody knife fights, gunshots, arson, and even murder. California had no true legal system until 1850, and even after law and order was given some clout, the brutish and reckless residents of the new state could often commit nefarious acts with impunity and no fear of retribution. Miners and townsfolk were justifiably frightened of what the night might bring, what dangers lurked in shadowy corners. Albert Evans, a Gold Rush journalist originally from New Hampshire, wrote that the allegorical symbol of Lady Justice had taken a beating during the earliest days of the rush: "If the plain truth must be told, Dame Justice . . . traveled so long in devious and crooked ways that she became permanently disabled, and never fully recovered the free use of all her faculties, having a cast in her unbandaged eyes, and a peculiar shuffling limp in her gait as she walks." The system improved marginally with statehood, but justice remained elusive. Apprehension was a constant.

Something had to be done, and soon, but the agencies charged with public safety and protection were ineffectual and mistrusted. The *San Francisco Evening Picayune* in August 1850 aptly described the bitter public sentiment toward civic leadership:

> There is scarce an officer entrusted with the execution of our state government, scarce a legislator chosen to frame the laws . . . , scarce a judicial officer from the bench of the Supreme Court down to the clerk of a village justice of the peace, scarce a functionary belonging to the municipal administration of our cities and incorporated towns who has not entered upon his duties and responsibilities as the means of making money.

Historian Hubert Howe Bancroft was more succinct and biting in his 1887 analysis: "Murderers were our congressmen, and shameless debauchers our senators. Our legislators were representatives of the sediment of society, and not worthy citizens."

For William Perkins, a merchant in Sonora, Tuolumne County, the solution seemed straightforward, but it raised the ominous specter of vigilantism:

> In a country like this, it is ridiculous to be hampered with ideas applicable to communities and societies already civilized. . . . The only means of purging this country of the dangerous villains who infest every town and settlement is to hang them. . . . We want a court that will try, sentence, and execute a man on the same day. By no other means will we be able to rid this country of the thousands of ruffians who now commit crimes with impunity.

This viewpoint was manifest throughout Gold Country, from San Francisco to Lake Tahoe, from Weaverville to Mariposa. So-called vigilance committees blossomed like flowers in spring. The law was taken into the hands of incensed citizens, who usurped inadequate legal authority and employed the stockade, the lash, and the rope throughout the region. Anarchy reigned and chaos was king.

In Sonora, the tension was exacerbated by the controversial and unpopular Foreign Miners' Tax, passed by the California State Legislature in April 1850. This tax placed a monthly levy of $20 on foreign miners, and its impact in multinational, multiethnic Sonora was particularly keenly felt. Widespread resistance to the law led to angry confrontations between foreigners and Anglo-Americans, both in the streets and in the diggings, and the results ranged from destruction of property to murder. The violence transformed Sonora from a respectable camp into a hellhole.

On July 3, 1850, William Perkins noted in his diary, "We have had three or four fresh murders. They are becoming so common that I hardly think of putting them down in my Journal. The times are becoming dangerous. . . . Scarcely a day passes but some murderous atrocity is committed."

In Sonora, the "lynch court" held sway. In July 1851, Tuolumne County Sheriff George Work arrested a notorious criminal named David Hill. For the prisoner's own safety, Work planned to transfer the man to Sonora at night to escape detection from those who might wish him harm, but before the transfer could occur, someone leaked the plan. The local vigilance committee determined that the criminal must die *post haste*, and on a moonlit night, dozens of committee members lined the town streets awaiting the arrival of the sheriff and his prisoner. Upon entering Sonora, Hill attempted to escape but was quickly captured by the waiting crowd. Hundreds gathered as the chant "We have got him! We have got him!" echoed through the darkened city.

William Perkins witnessed the events that followed. The captured man was led to a large oak tree. A circle of men ringed the hanging tree, "each one having a large revolver in his hand." The noose was tightened around the criminal's neck. According to Perkins, "a couple of score of men" grabbed the rope and, when the order was given, yanked, and "in a moment, a black object shot up, and loomed in obscure relief against the bright starlit heavens. . . . Death . . . was almost simultaneous with the act of running the man up."

Perkins's blood was high, his pulse was racing, and he was gratified that the murderer had been "treated in the same manner" in which most countries destroy "mad dogs, wolves, and other noxious animals, not only without compunction but with the satisfaction attending the performance of a duty."

A few weeks later, in bustling Sacramento City, another vigilance committee struck. The *Sacramento News-Letter* labeled the violent tableau that ensued a "theatre of scenes which we hope will never be repeated." The incident began in downtown Sacramento when James Wilson was mugged at 3 P.M. on July 9, 1851. Within hours, four perpetrators were arrested and charged with grand larceny—a capital offense. On July 12, they pled "not guilty" in court, and the trial was postponed until July 15 to allow the accused time to prepare their defenses. The vigilance committee would have none of that. A crowd stormed the courtroom and gave the presiding judge an order accompanied by the brandishing of loaded firearms: the trial of the first defendant, William Heppard, was to proceed within one hour. The committee appointed a posse to guard the other prisoners. When the hastily ordered affair began, Heppard was found guilty and sentenced to death. The remaining defendants were tried on July 15, as originally planned. One named Carrothers received a sentence of ten years, and the other two, James Hamilton and John McDermott, were sentenced to hang. The public executions of Heppard, Hamilton, and McDermott were slated for August 22, 1851.

On the morning of August 22, according to the *Sacramento Daily Union* account, "teams, horsemen, and pedestrians were seen pouring into the city from every direction, and at an early hour the city was crowded with miners and strangers from the country, who had come to witness the execution of the three culprits." The throng gathered around a scaffold that had been erected next to an old sycamore tree at Sixth and O Streets. The *Union* reported, "Business was entirely suspended, the streets were deserted—the city was at the scaffold. Every house, shed or elevation from which a view of the scaffold could be obtained, was crowded with human beings." The newspaper estimated the gathering at seven to eight thousand.

As the throng gathered, news arrived that Governor John McDougall, concerned over the forced trial of Heppard, had granted the convicted criminal a reprieve of his execution until September 19. Though angered, the vigilance committee agreed to leave Heppard in jail under the custody of the sheriff, even as Hamilton and McDermott were brought forth for hanging. But the unruly assembly was livid and hollered for Heppard to be hung immediately as well, crying, "Hang

THE CALIFORNIA VIGILANTES EXECUTING THE ORDERS OF JUDGE LYNCH.

The often lawless and violent atmosphere of the Gold Rush sometimes gave rise to vigilante justice. "The California Vigilantes Executing the Orders of Judge Lynch," from *Conquering the Wilderness,* by Frank Triplett (Chicago: Werner, 1885). Courtesy of the California State Library, Sacramento; California History Section.

---

the rascal!" "Hang the scoundrel!" "Bring him here!" "Let him hang, too!" The mob surged to the jail and snatched Heppard as sheriff's deputies watched.

Hamilton and McDermott were hanged legally under the supervision of Sheriff Benjamin McCullough. And then the muffled drums of the vigilante guards sounded and Heppard was escorted to the gallows. The sheriff retired from the scene, unable to change the pending outcome. William Heppard was given the opportunity to address the restless horde, and he launched into a profanity-laden attack on high-ranking officials, prominent citizens, and the vigilantes, accusing them of numerous improprieties. His remarks, characterized by the *Sacramento Union* as "disconnected and incoherent," were met with derision. He was hanged and his body was left suspended for twenty minutes.

That night, the multitude expressed their displeasure with Governor McDougall for issuing a reprieve to Heppard. An effigy of the governor was burned before "an immense concourse," the *Sacramento Union* noted, and "the utmost satisfaction was exhibited by the crowd" as the likeness was "undergoing martyrdom."

A few days later, the *Sacramento News-Letter* published the jailhouse confession of William Heppard. In it he stated that his only regret was that he "should be hung for so small an offense, and not one of his big transactions."

GOING UP THE FLUME

THE HANGING OF JOHN BARCLAY

On the sleepy afternoon of October 10, 1855, John Smith was thirsty. A miner from nearby Knickerbocker Flat, Smith was already tipsy when he entered Martha's Saloon at the corner of Jackson and Main Street in downtown Columbia. He stumbled into the barroom and, according to four witnesses, accidentally toppled—or, more likely, intentionally tossed—and shattered a pitcher from the bar counter. Martha Carlos Barclay, the bar's owner and a former prostitute, entered from an adjoining room and angrily inquired who had caused the damage. Fiery words were exchanged when Smith abruptly seized Martha and roughly tossed her onto a chair. John Barclay, Martha's new husband, heard the commotion and rushed into the saloon to confront Smith. Barclay drew his pistol and fired. John Smith was fatally wounded and died within minutes.

What happened next epitomized the violent and precipitous justice of the California Gold Rush.

John Barclay was a native of New York who had lived for several years in Chinese Camp, about fifteen miles from Columbia, working a profitable claim. While in Chinese Camp, he met Martha Carlos, who was well known in the region. Martha, believed to have been in her early twenties at that time, had been operating the Lone Star saloon, brothel, and boarding house in Columbia for about a year, and everyone knew the place as Martha's Saloon. In August 1855, the *Columbia*

Gazette reported that Martha "went to Chinese Camp, taking with the inmates of her house, and a business was opened there. While in Chinese Camp, she became enamored with Barclay, and the sentiment being duly returned, marriage followed!" With a promise that Martha would reform, the couple married on September 15, 1855. Following the ceremony, John accompanied Martha to Columbia and reopened Martha's Saloon.

Twenty-five days later, with gun smoke still hovering and Smith's bloody remains sprawled on the wooden plank floor, Barclay was arrested and hustled to the Columbia jailhouse. Within minutes, a crowd gathered, thirsting for swift justice. John W. Coffroth, a lawyer and former California state senator who lived only one hundred feet from the saloon, whipped the growing crowd into a frenzy. This incident was so heinous, Coffroth exclaimed, that speedy vengeance must be exacted. The victim was a friend, he continued, and Coffroth urged the mob to take the law into their own hands. He had been elected to enact laws, the senator bellowed, but this case called for suspension of due process and direct remedy via the harsh justice of Judge Lynch.

The angry mob stormed the jail and disarmed the town marshal and the jailers. A keg of black powder was placed at the foot of the jail in the event it was needed to blast John Barclay out of confinement, but it was not necessary. Men hacked away at the iron bars and cell walls with crowbars, sledgehammers, and axes until the doors gave way.

Barclay was dragged out of the building and, when an opportunity briefly arose, he made an attempt to run. He was quickly apprehended, and calls for an instant trial rang through the streets of Columbia. By acclamation, John Heckendorn, editor of the *Columbia Clipper* newspaper, was chosen as judge, and a jury was hastily selected.

The teeming, menacing crowd marched a few hundred yards to the flume of the Tuolumne County Water Company and the trial began. "Judge" Heckendorn presided and John Coffroth was appointed prosecuting attorney. John Oxley, state assemblyman for the district, was given the unenviable task of defending John Barclay.

The four eyewitnesses testified to the circumstances of the shooting. Any attempt by Defense Attorney Oxley to cross-examine was shouted down. Oxley was not allowed to call witnesses or character references in support of John Barclay. The result was a foregone conclusion and everyone, including Barclay, knew it.

In his final remarks to the jury, John W. Coffroth seized the moment to present an emotional and dramatic plea for justice for his friend John Smith:

> Gentlemen, I have but little to say. You all knew the deceased, and knew that he was honest, good, and high-minded. You have heard all the testimony and know the witnesses; they have all lived long among you. . . . The only question to ask is, Who is the murdered man, who the murderer? If you are satisfied that the prisoner shot Smith, then it is your duty to declare it, and it is your duty to declare the penalty. . . . [As to forgiveness], there is a higher Court to ask for mercy.

The rich and colorful Tuolumne County mining camp of Columbia was the location of one of the most dramatic and horrifying instances of vigilantism during the Gold Rush. "Columbia, Tuolumne County, 1855," lithograph by Kuchel and Dressel. Courtesy of the California State Library, Sacramento; California History Section.

A trembling John Barclay, knowing the end was near, begged Coffroth to ask the jury for some time to put his affairs in order. His plea was ignored.

John Oxley rose to speak on Barclay's behalf. His words were swamped by a rising crescendo of catcalls and angry denunciations.

"I shall be brief," Oxley began. "Consider well, gentlemen, what you are about to do. Let to-morrow bear favorably upon the acts of this night."

These words were drowned by yells of "Enough! Enough!"

Oxley soldiered on, "Will you not sustain the laws? Will it not be better that the just laws of the land should take their course?"

"No! No!" the mob cried, "Up with him! Damn the laws!"

The defense attorney continued, raising his voice to reach above the swelling chorus, "Let him be confined to jail. Consider your course, and the great responsibility that you assume. Give time for reflection. Let calmness have time to come in. Do not, after you have taken this man's life, find that it is too late to do justice."

"Enough!" "Hang him!"

Judge Heckendorn urged Oxley to end his remarks as the increasingly enraged throng yelled "Drag him up!" "He gave Smith no time!" "Hell shall not save him!"

John Oxley gave up.

As Heckendorn gave final instructions to the jury, James Stewart, the sheriff of Tuolumne County, arrived. Stewart had been sheriff for just over a month and he was about to be thrust into the eye of a hurricane.

Sheriff Stewart demanded that Barclay be released to him. The mob angrily declined. Someone grabbed Stewart by the throat, tossing him violently to the ground. The crowd screamed "Sheriff! Sheriff!" as they carried Stewart and Barclay in opposite directions. Barclay was taken to the flume and a rope strung around his neck. The sheriff broke loose and rushed toward Barclay, again insisting that the prisoner be released into his custody. From somewhere, Sheriff Stewart produced a knife and tried to sever Barclay's noose. During this futile attempt, Stewart was pistol-whipped and beaten by the mob, and his clothes were torn.

He was saved by a solitary soul who courageously brandished a bowie knife and repelled the attackers.

The moment approached for Barclay to be hanged, or to "go up the flume," as the saying went. With the hangman's rope looped over the flume bracing, Barclay's executioners began hoisting the prisoner upward. The unruly crowd failed to pin Barclay's arms, however, and the desperate victim grabbed the rope with both hands. To break his hold, the hangmen repeatedly and violently jerked the rope up and down. As they did so, they cried out to Barclay to let go. His strength depleted, John Barclay released the rope and his hands fell. There was a final convulsive spasm and he was dead. In 1882, historian Herbert Lang described the mob's reaction, based upon eyewitness accounts: "The spectacle was said to be truly horrifying; a human form, hanging by the neck, in mid-air; a vast throng of men, shouting, yelling and jumping; while the red and lurid glare of torches and bonfires sent a horrid flash upon the terrible scene."

Six hours had elapsed between the shooting of John Smith and the execution of John Barclay.

The shocking brutality of the incident was met with widespread condemnation. Nine days later, the *Sacramento Union* opined:

> Our character abroad is made up from such lawless acts by a mob of excited men as this one in Columbia, and until an end is put to such acts of mob violence, we shall be looked upon as little better than the savages in Patagonia. In this case there was no justification for a resort to the Lynch code. The only end gained was the gratification of a fierce spirit of revenge. . . . It was a sad day for Tuolumne, and a sadder one for the reputation of the people of the State.

Even in Columbia, thoughtful residents denounced the horrific episode. Three days after the lynching, the editors of the *Columbia Gazette* offered this coda:

> We are not, we never have been, and never will be, the advocates of mob law, under any form, or for any end whatever. . . . We ask any man who looked on calmly

> (if any could do so) what chance any one stood for justice with the throng on Wednesday. . . . It pains us to record these occurrences as having taken place in Columbia, and we would gladly omit them, but our duty as journalists compels us to publish them; they are a blot upon our town. We trust in heaven that this may be the last time we shall have to perform so disagreeable a task; and we congratulate all those who have had no participation in these lamentable occurrences.

The Columbia hanging of John Barclay proved to be the beginning of the end for lynch law as applied to Anglo-Americans in California, but mob violence against Native Californians and Chinese immigrants would continue for decades.

"AWFUL CALAMITY"

THE EXPLOSION OF THE STEAMER *BELLE*

Steamboats were critical cogs in the massive commercial machine that powered the Gold Rush. As the goldfields prospered and grew from pockets of isolated prospectors into mining camps and then towns, consumer goods went from being niceties to necessities. The question was how to deliver the merchandise. The San Francisco Bay, the Sacramento and San Joaquin Rivers, and every navigable tributary and slough were the conduits of commerce linked to the spider web of trails and roads in Gold Country by stagecoaches, freight drayage, pack mules, express riders, and even dogsleds.

Like so much else in California, the specialized steamboats had to be imported at great expense. But where there is great profit, there are investors, and several entrepreneurs placed orders for shallow draft riverboats from companies on the Mississippi River, and even as far away as Maine. John David Borthwick observed that their safe arrival was nothing short of a miracle. It was a stunning accomplishment, he wrote, that "these fragile-looking fabrics" could be delivered undamaged "over the seventeen thousand miles of stormy ocean . . . [;] one could not help feeling a degree of admiration and respect for the daring and skill of the men by whom such perilous undertakings had been accomplished." The steamboats earned immediate approbation in San Francisco and were soon plying inland shipping routes. By 1852, thousands of tons of cargo were being delivered to Sacramento and other embarcaderos in the interior.

In the beginning, steamer demand far outstripped supply, and profits were high. In 1850, freight shipped from San Francisco to Sacramento fetched $40 a ton. Cabin passengers plunked down $30 per person for the same trip. Success spawned competition both in transport and construction, and soon shipyards blossomed in San Francisco and Sacramento and produced steamers designed especially for California conditions. In 1853, these California steamboats joined a growing cadre of crafts aggressively competing for business. Increased shipping capacity resulted in a precipitous fall in rates and revenue. By 1854, on the San Francisco–Sacramento route, freight shipped for $1 a ton, and cabin passenger fares dropped to $3 per person, with deck passage a mere twenty-five cents. Rate wars characterized the cutthroat competition, as did physical confrontation. Rival companies scuttled boats and rammed their adversaries. There were reports of crews taking potshots at their challengers. Bribery was common, and lower profits led to deficient upkeep. Bored passengers sought thrills and urged steamboat captains to ever greater speeds or to engage in risky races. Increased velocity required more steam, and more steam combined with poorly maintained boilers inevitably led to explosions. There had been accidents before, but as the 1850s progressed, disasters were increasingly common.

In 1854, with rates plummeting and damage increasing, steamboat owners created an umbrella company to end the catastrophic disputes and form a cartel. They called themselves the California Steam Navigation Company, but they were more commonly referred to as "The Combination." Pooling their resources, establishing a fixed rate schedule for freight and passengers, and removing boats from the river, the Combination stabilized the steamboat enterprise and boosted profits. It also generated vitriolic outrage from newspapers, the public, and vulnerable merchants, who decried what some christened the "monster monopoly." The Combination continued to consolidate its power through new freight-rate manipulations, more ramming, intensified bribery, and, when that failed, simply buying their competitors and forcing them out of business. By 1858, most merchants had capitulated and the powerful steamer alliance dominated river traffic. Meanwhile, a group of fledgling steamer companies formed a separate organization known as "The Opposition." Prominent figures supported efforts to break the monopoly, but it was no use. Shopkeepers and customers cursed delays caused by deadly accidents, silted watercourses, and inflated prices, but there was little they could do, and, at the end of the day, the goods they needed to ship were necessary to feed the voracious appetites of the burgeoning goldfields.

But, on February 5, 1856, none of this mattered. The steamboat *Belle*, a Combination vessel, had left Sacramento at around 6:45 that morning, bound for Red Bluff. Eleven miles north of the city, passenger Thomas McAlphin strolled toward the mess for breakfast. On the passenger deck he noticed a man sitting close to a potbellied stove, one leg propped on either side to keep warm on the chilly morning. Suddenly, the steamboat shuddered and rumbled as a huge explosion ripped through the *Belle*. The potbellied stove shot like a rocket through the roof, ripping through the hurricane deck, and landed overboard with a massive splash. When it hit the Sacramento River, the stove was in the company of floating debris, dead bodies, and severed limbs. The man seated at the stove was stunned but uninjured.

Aboard the vessel were forty passengers, twenty crew, and forty-five tons of cargo, including groceries, dry goods, and "treasure," believed to be a cache of gold coins. Even after extensive investigation, it was never learned why the boiler exploded; it wasn't due to excess steam, since earlier that morning Captain Charles Houston had delayed the ship's launch due to heavy fog and ordered his engineer, Washington Elricks, to proceed cautiously under reduced steam.

The force of the explosion had snapped the boat's keel, shredded the main deck, torn through the main salon, and annihilated the pilothouse. Captain Houston was killed instantly, as were the other deck officers. Several other members of the crew perished in the explosion, including fireman William Green. Only half of Green's body was found. One of the two large paddlewheels detached from the steamer and listed alarmingly in the water next to the shattered vessel.

The explosion created a halo of blood and debris, and within moments, as the *Sacramento Daily Union* reported, there was a gruesome jumble: "From one to the other end of the boat, in inextricable

Steamboats were a vital cog in the vibrant Gold Rush economy, but they could also be hazardous and deadly. "Californie. Vue de la Ville de Sacramento, 1850," illustration by Urbano Lopez, lithograph by J. Hannin (Paris: 1850). Courtesy of the California State Library, Sacramento; California History Section.

confusion, were mingled cots, chairs, tables, provisions and fragments of the machinery; and, horrible to relate, clots of brains were everywhere to be seen and bespattered on the doors of the state-room, and on almost every plank and timber, were observable the blood of the unfortunate victims of this calamity."

Soon, observers on shore raced to assist survivors and remove mutilated remains from the river with grappling hooks. Doctor Ruduck, a farmer, had heard the explosion and rushed to the riverbank. He spied survivors clinging to drifting wreckage and, using his own rowboat, was the first on the scene. As he raced to these survivors, Ruduck saw two victims sink beneath the water, but he was able to rescue five others. Several ships arrived to help with the search, recovery, and evacuation efforts. One boat in particular was singled out for its heroic efforts. It was a small steamer named the *General Reddington*. One account recalled that the *General Reddington* was "everywhere, . . . working, commanding, consoling and doing all that man could do to succor the wounded and recover the dead." Witnesses recalled ship documents, such as bills of lading, driven skyward by the force of the blast, fluttering like butterflies distant from the river.

An estimated thirty people died or were seriously wounded. Some bodies were never recovered. Among the lesser injuries were shattered limbs, blistering and scalding, and numerous cuts and bruises. Among the injured passengers was John Bidwell, a prominent landowner, politician, and founder of Chico. Bidwell, who later served in the California State Senate and the United States Congress and ran as a third-party candidate for California governor and for president of the United States, fractured his skull in the *Belle* explosion.

There were thrilling stories of survival, albeit often mixed with grim postscripts. William Mix was in the main salon having a conversation with a friend when the ship exploded and an iron missile spiraled through the cabin and struck his friend in the head, killing him and splattering his brains over Mix's coat. When Mix returned to what remained of his cabin, he discovered that shards of boiler had rocketed through the room. Unharmed was the caged canary he had left on the table before leaving for breakfast. The table had disintegrated, but the cage was only dented in the fall. A few days later, the *Sacramento Daily Union* happily announced, "The bird is now singing as sweetly as ever at the Orleans [Hotel], the accident having merely disturbed its plumage."

Passenger James Powell had just left the breakfast table when the blast had rudely lifted him off his feet and propelled him through the lattice door of the washroom. He suffered only minor scrapes.

The *Belle*'s bartender had been serving in the center of a circle of patrons when disaster struck. All the surrounding customers died or were severely injured by the explosion. Metal shrapnel neatly sliced a portion of the rim from the bartender's hat and tore a long, jagged gash in one of his leather boots, but he was unharmed. James Hyland, the *Belle*'s steward, was strolling through the main cabin when the explosion wildly sprayed deadly iron shards through the room. The metal projectiles cut his hat in two, but Hyland was untouched.

On February 6, 1856, when the death toll was still unknown, the *Sacramento Daily Union* conjectured, "We fear that a great proportion [of the missing] are no longer in the land of the living, and there is little probability that their names will all be recorded, save in the registry of Heaven. This deplorable tragedy . . . has cast a deep gloom over our city."

The name of at least one victim was etched somewhere more accessible than Heaven. Twenty-three-year old Leonidas Taylor, the *Belle*'s clerk that day, was never seen or heard from after the explosion, and his body was never recovered, but in October 1856, a marble obelisk to commemorate Leonidas was erected by his brothers on a lonely country road alongside the Sacramento River near the site of the *Belle* tragedy. It still stands today. The inscription reads:

> Erected to the memory of Leonidas Taylor, born in the City of Philadelphia on the 3rd of July, 1832. He grew to manhood in the City of Saint Louis and was killed by explosion of the steamer *Belle* opposite this spot on the 5th of February, 1856. His body was never found. Far distant from those who loved him, the waters of the Sacramento will roll over him till that day when the sea shall give up its dead.

The steamboat *Belle* calamity did not mark the end of steamer travel on California rivers, nor did it mark the ascent of stringent safety regulations. Accidents and explosions continued to happen. On August 25, 1861, the boiler of the *J. A. McClelland* exploded, killing fifteen. The engine of the *Washoe* blew up on September 5, 1864, while the steamboat was racing the paddlewheel steamer *Chrysopolis*. Nearly half of the 175 passengers on the *Washoe* perished, with more than forty survivors severely injured. On October 12, 1865, the boiler of the *Yosemite* exploded on the Sacramento River, killing fifty-five and injuring dozens more. As other forms of transportation took their place, the number of paddlewheel steamboats dwindled every year, although they continued to ply California inland waterways until the early 1940s.

THE TERRITORY OF COLORADO

THE PICO ACT OF 1859

The vast majority of California's new arrivals in the years after 1849 landed in Gold Country and the central Sierra Nevada, and these regions saw the brunt of the state's upheaval in its first decade. But sometimes a lack of change can be felt as keenly as change itself. In the southern and far northern portions of the state, which saw their clout plummet as rapidly as their share of the population, political and social distress quickly followed. Feeling underrepresented, ignored, and overtaxed, the residents bracketing the gold regions often supported schemes to sever their localities from the state starting as early as the California Constitutional Convention of 1849.

In 1859, Andrés Pico—a member of the Californio elite, a brigadier general in the California Militia, and a prominent state legislator from the Los Angeles area—introduced a resolution to establish a new and separate territory carved from six counties of Southern California located below the 36th parallel. Refined in committee, the legislation, titled "An Act Granting the Consent of the Legislature to the Formation of a Different Government for the Southern Counties of the State," called the new province the "Territory of Colorado" and would require not only approval by the state legislature but a plebiscite by those living in the area. If two-thirds of those in the affected area voted yes, the proposal would be sent to Congress and the president for endorsement. At the same time, delegates from the far northern counties of California introduced legislation to form the "State of Klamath" in their region for the same reasons.

Andrés Pico: a brigadier general in the California Militia and the author of legislation to split Gold Rush California into two states. Portrait from University of California, Berkeley; Bancroft Library; Pico, Andrés–POR3.

Pico's bill passed the California Assembly on March 25, 1859, by a vote of 33 to 25; it squeaked by 15 to 12 in the State Senate on April 14, and Governor John Weller signed the legislation a few days later. The passage sparked a lively debate, with most arguments against centered on the act's constitutionality.

As provided for in the legislation, the question of whether to form a Territory of Colorado in the southern counties was put to a vote in the autumn of 1859 in the area directly involved. The final tally was 2,457 to 828 for separation—75 percent of voters approved of the idea, far in excess of the two-thirds needed.

In January 1860, former governor Milton Latham, just selected to replace the recently deceased David Broderick as the United States senator from California, attempted to shepherd the bill through Congress. Latham wrote an appeal to President James Buchanan:

> [The southern counties are dissatisfied with] the expenses of a State Government. . . . They complain that the taxes upon their land are ruinous—entirely disproportioned to the taxes collected in the mining regions; that the policy of the State . . . having been to exempt mining claims from taxation, and the mining population being migratory in character . . . and contributing but little to the State revenue in proportion to their population, they are unjustly burdened; and that there is no remedy save in separation from the other portion of the State. In short, that the union of southern and northern California is unnatural.

But Latham's advocacy fell on deaf ears. In 1860, the nation was on the brink of civil war and rebellious Southern states were threatening secession. While introduced in congressional committees, the Pico Act never came to a vote, and once the Civil War erupted, the legislation was forgotten.

The Pico Act of 1859 was not the first attempt at splitting California into separate states or territories, nor would it be the last. More than two hundred such proposals have been floated since the Gold Rush, including the current movement to establish a "State of Jefferson" in the upper reaches of Northern California.

THE FINAL ACCOUNTING
DEATH IN THE GOLDFIELDS

The Grim Reaper was a constant companion during the Gold Rush. Accidents, disease, murder, natural disasters, mob violence, and the inevitable pageant of the ages took a heavy toll during the era. Some estimates indicate that 20 percent of all forty-niners died within six months of reaching California, and a significant number of those who undertook the journey died along the way. All recognized the potential dangers and the manifest insecurity of the adventure, but perishing in a faraway land among a host of strangers was often not considered a possibility by the largely young and youthful participants. The sorrowful tales of those who had spent their last pennies on an epic journey, covering thousands of miles to the goldfields, and who then expired soon after arrival are grim entries in journals of the time.

And yet, surprisingly few modern Gold Rush histories address death at any length, instead focusing on the overland journey, rollicking adventures in the mining camps, and the churning tumult of socioeconomic affairs. Some Gold Rush books do not mention death at all. Of course, death awaits every one of us and mortality has been a disquieting topic throughout human history, but in this land built on fantasy, the final accounting seemed somehow amplified and more emotionally penetrating. Even though most who came to California suffered failure, they had all come brimming with hope, anticipating a glimmering end to poverty or a new beginning, and thus, perhaps, death in the California Gold Rush took on a unique quality, one that spoke to the outlook of the argonauts. After all they had endured in pursuit of the golden dream, death felt a particularly harsh affront, an especially cruel insult.

As John David Borthwick remarked, California Gold Rush society was "a picture of universal human nature boiling over." The experience was intense, concentrated, and many shared a communal history of sacrifice and hardship. A death in the goldfields was not only a lost life but the end of a dream. Many felt these deaths personally, even if they did not know the deceased, for they were part and parcel of the same dream. The journals and letters of the era powerfully communicate these feelings.

Let us start with accounts from the trail to California. Or even earlier than that: as Alonzo Delano reported, some hopefuls passed even before the journey had commenced. As Delano's party prepared to depart St. Joseph, Missouri, a young member of the wagon company took ill, rapidly deteriorated, and died before he had traveled even an inch toward California. Delano recounted that, in the hills overlooking the Missouri River, the company selected a gravesite:

> A bright green sward spread over the gentle slope, and under a cluster of trees his grave was dug. . . . A procession was formed by all the passengers, [and] proceeded to the grave, where an intimate friend of the deceased read the Episcopal burial service, throughout which there was a drizzling rain. . . . How little can we foresee our own destiny! Instead of turning up the golden sands of the Sacramento, the spade of the adventurer was first used to bury the remains of a companion and friend.

Trailside ceremonies could be stark and heartrending. Catherine Haun, a young newlywed from Iowa, had traveled with her lawyer husband to California in the pioneering days of 1849. She described the loss of a Mrs. Lamore and her infant during childbirth on a desolate desert expanse, deaths that left behind an inconsolable husband and two motherless little girls:

> We halted a day to bury her and the infant that had lived but an hour, in this weird, lonely spot on God's footstool away apparently from everywhere and everybody.
>
> The bodies were wrapped together in a bedcomforter and wound, quite mummyfied with a few yards of string that we made by tying together torn strips of a cotton dress skirt. . . . Every heart was touched and eyes full of tears as we lowered the body, coffinless, into the grave. There was no tombstone—why should there be—the poor husband and orphans could never hope to revisit the grave and to the world it was just one of the many hundreds that marked the trail of the argonaut.

If a young miner did pause to consider the potential circumstances of his demise in California, there were several likely scenarios: disease, some variety of mining accident, or maybe even a drunken fight in a raucous camp saloon. But other deaths were extremely unlikely and wouldn't have entered their minds. William Shaw, an English gold seeker, vividly remembered a night near Tuleberg (later known as Stockton), when one such unlikely circumstance occurred. Shaw was looking for a place to sleep and, as he wrote, "to my great joy a space under a wagon was unoccupied, so lying down on some rotten wood and rushes, I was just falling asleep, when an exclamation of pain and horror from an adjacent sleeper aroused me." Shaw had heard the moaning of a "hale gigantic man of about 30, who had been stung by a venomous insect, . . . the sting of which he knew to be mortal." The ailing man was soaked with sweat and his body shuddered from tremors. "Various remedies were proposed," Shaw recalled, "but he shook his head: 'No,' said he, 'die I must.'" As the venom took hold, the stricken man drank copious amounts of brandy, philosophized about his mortality, and reminisced about his experiences as a veteran of the Mexican War. Shaw watched in alarm as "violent spasms soon came on, and he shouted for more liquor; his features seen by the lurid light of the fire were horrible to contemplate, and it was not without violent struggles that he gave up the ghost." Deeply shaken, Shaw sought other sleeping arrangements that night.

Many argonauts had mortgaged their futures to seek the elusive golden bounty, and most had promised family and friends that they would return home rich beyond imagining. When they failed, as the majority did, the sting and shame of failure led some to contemplate the ultimate, permanent end to their anguish. As Hinton Rowan Helper surmised in his caustic 1855 book *The Land of Gold: Reality Versus Fiction,* "In California, where projects are pursued with a recklessness elsewhere unknown, the losses are on a gigantic scale. Disappointments . . . have the keenness of those of the beaten gambler, to whom defeat is irretrievable ruin. What wonder, then, that suicides are so common in that unhappy country?" Reports of suicides or attempted suicides became routine, yet another sadly familiar aspect of daily life. Edmund Booth, a forty-niner from Iowa, tersely described the suicide of his mining partner in Sonora in an 1852 letter to his own wife, Mary Ann: "[He] cut his throat with a razor—dead of course." Heinrich Lienhard, an early resident of Sutter's Fort, wrote of an incident in which John Sutter had angrily berated his daughter, causing such emotional distress that "she tried to commit suicide by cutting an artery on her wrist with a knife. I was even told that Sutter offered her a pistol, telling her she could end her life quicker that way."

In the initial days of the Gold Rush, funerals were frequently spare and improvised. In September 1851, Louise Amelia Knapp Smith Clappe (Dame Shirley) wrote of a makeshift funeral for one of the four women residing in the rugged settlement of Rich Bar on the Feather River. The woman, "poor Mrs. B," had died of peritonitis.

> Her funeral took place at ten this morning. . . . On a board, supported by two butter-tubs, was extended the body of the dead woman, covered with a sheet. By its side stood the coffin, of unstained pine, lined with white cambric. . . . An extempore prayer was made, filled with all the peculiarities usual to that style of petition. Ah, how different from the soothing verses of the glorious burial service of the church! . . .
>
> As the procession started for the hillside graveyard, a dark cloth cover, borrowed from a neighboring monte-table, was flung over the coffin. . . . Should I die to-morrow,

The Gold Rush was exhilarating, but life in California could also be perilous, and the Land of Gold became the final resting place for many hopeful miners. "The Departure," *Hutchings' Illustrated California Magazine* 5, no. 3 (September 1860). Courtesy of the California State Library, Sacramento; California History Section.

I should be marshaled to my mountain-grave beneath the same monte-table-cover pall which shrouded the coffin of poor Mrs. B.

I almost forgot to tell you how painfully the feelings of the assembly were shocked by the sound of the nails (there being no screws at any of the shops) driven with a hammer into the coffin while closing it. It seemed as if it *must* disturb the pale sleeper within.

Sometimes the improvised nature of miners' funerals lent them an air of the absurd. In 1848, James Carson, a miner who had arrived in California prior to the rush, attended a funeral on the South Fork of the Stanislaus River. A popular gold seeker named George had died, and as Carson recalled, the camp "determined to give him as respectable a funeral as circumstances would permit." An unforgettable diggings denizen called "The Parson," who was once a "powerful preacher" back East but had fallen on hard times and become a slave to alcohol, was called upon to officiate. The ceremony commenced as "many a drink went down to the repose of the soul departed." The Parson took more than his share, and when he began his eulogy, Carson remembered, the minister was "somewhat muddled." The dearly departed was deposited in his newly dug grave on the river bank, and the Parson offered a prayer, read a long passage from the Bible, and then encouraged everyone to join him in a hymn. There were no hymnals and, unfortunately, only the Parson knew the tune and lyrics, which he bellowed out, until he stopped suddenly, having forgotten the words. No matter, Carson reported, as "he coolly informed us that the Lord had obliterated from his memory the balance of that solemn Psalm, but we would go to prayer." As those in attendance bowed their heads, they "commenced examining the dirt that had been thrown up and found it to be (as they expressed it) 'Lousy with gold.'" This caused a commotion at the graveside. The Parson ceased praying, looked down at the grave, and excitedly he exclaimed, "Gold! by God!—and the richest kind o' diggins!—the very dirt we have been looking for!" He then raised his hand in benediction, announced the congregation dismissed, and raced off to find his pick and gold pan. Poor George was removed from his grave and reburied, as James Carson noted, "high up the mountain's side." The first gravesite was then panned and proved a very rich hole indeed.

George was fortunate to have friends—if a dead man could be called fortunate. Many accounts of the period sorrowfully recollect the demise of unidentified, nameless miners in rude camps separated by thousands of miles from the only people who could recognize or remember the deceased. Many of the families never learned the fate of those who had left. Pennsylvania forty-niner Daniel Woods recalled the passing of a mysterious German sailor from "a violent attack of diarrhea." Woods wrote, "Nothing could be learned of him or his friends—even his name was unknown to us. We buried him deep in the sand, on the banks of the Tuolumne." Woods added that, as the burial services were performed, a neighboring, oblivious crowd "surrounded the gambling-table on the bar."

John Steele was a gold seeker from Wisconsin who spent three years in California beginning in 1850. He recounted perhaps the most startling of the anonymous Gold Rush deaths in his memoir *In Camp and Cabin*. It was January 30, 1851, on the Feather River. Steele was part of a mining party that was seeking paydirt along the watershed. A member of his company who was exploring upstream returned excitedly to their camp. He exclaimed, "I reckon somebody has struck it rich down there, and covered up their prospect hole so as to hide it." John and two others accompanied the scout to a meadow to uncover the hole—not to steal the claim but to gain a clue as to possible rich diggings nearby. Upon arrival, Steele found broken ground and a gruesome surprise:

> Spading away the soft earth to the depth of about three feet, we found—not a gold mine, but that which made us start back with horror—a blue shirt sleeve on the arm of a corpse. Gently the body was uncovered and raised to the surface; water was brought and, washing away the mire, disclosed the features of a young man, of probably twenty years; about five feet in height; dark brown hair; his only clothing a blue woolen shirt, dark brown pantaloons, and heavy boots.

They searched for some proof of identity in his shirt and trousers, but the "pockets were empty and there was nothing about him to reveal his name." After cleaning the dead body, Steele and his compatriots discovered wounds indicating that a bullet had passed through the victim's skull. They reckoned the young man had been murdered a few days earlier and "his body concealed in this wild glen." John Steele described the emotional impact the macabre encounter had upon his fellow miners: "Tears filled our eyes as we thought of his untimely fate, and that father, mother, brothers, and sisters may lovingly await his return until hope deferred makes the heart sick. The death-sealed lips could not reveal the name of the murderer to men, but there is a Witness who knows all about it, and sometime the criminal will stand at the judgment bar of God."

Steele's party gathered the remains and delivered them to Downieville, Sierra County, where the unfortunate was buried without ceremony or headstone. "Long afterward," John Steele remembered, "when passing [through Downieville], I made diligent inquiry, and learned that no knowledge of the man's name, friends, or home had been found. To use a phrase common among mountaineers, he had been 'rubbed out.'"

SOURCES

Preface (pages xix–xxii)

Clark, William B. *Gold Districts of California.* Bulletin 193. Sacramento: California Division of Mines and Geology, 1970.

Cutter, Donald C. "The Discovery of Gold in California." In *Geologic Guidebook along Highway 49, Sierran Gold Belt: The Mother Lode Country*, pp. 13–17. Bulletin 141. Prepared under the direction of Olaf P. Jenkins. Sacramento: California Division of Mines, 1948.

The Wings of the Future: Beginnings (pages 1–4)

Bancroft, Hubert Howe. *History of California.* Vol. 6, *1848–1859*. San Francisco: The History Company, 1888.

Holliday, J. S. *Rush for Riches*. Berkeley: University of California Press, 1999.

Mason, Richard Barnes. Mason to Roger Jones, Adjutant General, August 17, 1848. *Journal of the 31st Congress, House Executive Document 17*, Washington, D.C., January 24, 1850.

New Orleans Daily Picayune, November 23, 1848.

Patterson, Robert M. Patterson to Robert Walker, Secretary of the Treasury, December 11, 1848. *San Francisco Weekly Alta California*, March 12, 1849.

Polk, James K. *State of the Union Address*. December 5, 1848.

Watkins, T. H., and R. R. Olmsted. *Mirror of the Dream: An Illustrated History of San Francisco*. San Francisco: Scrimshaw Press, 1976.

Undisciplined Squads of Emotion: Motivations (pages 5–10)

Armstrong, John Elza. "Diary of John Elza Armstrong." In *The Buckeye Rovers in the Gold Rush: An Edition of Two Diaries*, edited by H. Lee Scamehorn, Edwin P. Banks, and Jamie Lytle-Webb. Athens: Ohio University Press, 1989.

Brannan, Ann Eliza (or Lisa). "A. L. Brannan Gold Rush Letter," September 2, 1848. California History Section (SMCII, box 14, folder 9), California State Library, Sacramento.

Chase, Nathan. Nathan Chase to Jane Chase, March 5, 1852. Nathan Chase Letters. Beinecke Library, Yale University Library.

DeWolfe, David. Quoted in *Gold Rush Diary: Being the Journal of Elisha Douglass Perkins on the Overland Trail in the Spring and Summer of 1849*, edited by Thomas D. Clark, 152. Lexington: University of Kentucky Press, 1967.

Holliday, J. S. *Rush for Riches*. Berkeley: University of California Press, 1999.

Mason, Richard Barnes. Mason to Roger Jones, Adjutant General, August 17, 1848. *Journal of the 31st Congress, House Executive Document 17*, Washington, D.C., January 24, 1850.

Monterey Californian, August 14, 1848.

Nash, Charles. Letter to the editor, July 10, 1850. *Marshall [MI] Statesman*, September 4, 1850.

New York Herald, December 10, 1848.

Newell, William W. *The Glories of a Dawning Age*. Syracuse, NY: Thomas S. Truair, 1853.

Pérez Rosales, Vicente. *Recuerdos del Pasado*. 1855. In *California Adventure*, by Vicente Pérez Rosales. Translated by Edwin S. Morby and Arturo Torres-Rioseco. San Francisco: Book Club of California, 1947.

Pierpont, James Lord. "The Returned Californian." Song, arranged by John P. Ordway. Boston: E. H. Wade, 1852. [Pierpont is better known today for his 1857 song "The One Horse Open Sleigh," commonly called "Jingle Bells."]

Royce, Sarah. *A Frontier Lady*. New Haven, CT: Yale University Press, 1932.

Swain, William. Swain to his mother, July 19, 1850. Swain Diary and Letters. Yale University Library. [Swain is extensively quoted in *The World Rushed In*, by J. S. Holliday. New York: Simon and Schuster, 1981.]

Wierzbicki, Felix. *California As It Is, and As It May Be*. San Francisco: Washington Bartlett, 1849.

The Consequence of Fearful Blindness: Twilight of the Californios (pages 11–17)

"An Act to Ascertain and Settle the Private Land Claims in the State of California." Approved March 3, 1851. *Statutes at Large, Public Acts of the 31st Congress of the United States, Session II (December 1850–March 3, 1851)*, Chapter 41, 631–34.

Avina, Rose. *Spanish and Mexican Land Grants in California*. New York: Arno Press, 1976.

Becker, Robert H. *Designs of the Land: Diseños of California Ranchos and Their Makers*. San Francisco: Book Club of California, 1969.

Emparan, Madie Brown. *The Vallejos of California*. San Francisco: Gleeson Library Associates, University of San Francisco, 1968.

Gates, Paul W. "The California Land Act of 1851." *California Historical Society Quarterly* 50, no. 4 (December 1971): 395–430.

Gwin, William. Speech to the United States Senate, January 8, 1851. Congressional Globe, 31st Congress, 2nd Session (1850–1851), 159.

Moraga Historical Society. "The Story of Moraga." In *Moraga's Pride: Rancho Laguna de Los Palos Colorados*. Moraga Historical Society, 1987.

Oakland Daily Transcript, October 17, 1877.

Perez, Cris. *Grants of Land in California Made by Spanish or Mexican Authorities*. Sacramento: Boundary Determination Office, Boundary Investigation Unit, California State Lands Commission, 1982.

Rosenus, Alan. *General Vallejo and the Advent of the Americans*. Berkeley: Heyday, 1995.

Vallejo, Mariano Guadalupe. *Recuerdos Históricos y Personales Tocante a la Alta California*. 5 vols. 1875. Original manuscript in the Bancroft Library, University of California, Berkeley.

"I Know It to Be Nothing Else": The Sad Tale of James Marshall (pages 18–24)

Bekeart, Philip B. "James Wilson Marshall, Discoverer of Gold." *Society of California Pioneers Quarterly* 1 (September 1924).

Dunbar, Edward E. *The Romance of the Age; or, The Discovery of Gold in California*. New York: D. Appleton, 1867.

Gillespie, Charles B. "Marshall's Own Account of the Gold Discovery." *Century Illustrated Monthly Magazine*, n.s. 41, no. 19 (February 1891): 537–38.

Marshall, James, and George Frederic Parsons. *The Life and Adventures of James W. Marshall, the Discoverer of Gold in California*. Sacramento: J. W. Marshall and W. Burke, 1870.

New York Herald. "Who Discovered Gold in California." June 27, 1849.

Placerville [CA] Mountain Democrat. "50th Anniversary of California Gold Discovery." January 24, 1898.

San Francisco Chronicle. "Trailing Marshall." February 16, 1908.

Statutes of California Passed in the Nineteenth Session of the Legislature, 1871–72. Sacramento: T. A. Springer, State Printer, 1872.

Sutter, John A. Personal reminiscences, as dictated to Hubert Howe Bancroft in 1876. Original manuscript in the Bancroft Library, University of California, Berkeley.

Williamson, R. S. "Report of Explorations in California for Railroad Routes." Vol. 5, part 1 of *Reports of Explorations and Survey to Ascertain the Most Practicable and Economical Route for a Railroad from the Mississippi River to the Pacific Ocean.* United States Senate, 1855. Washington, D.C.: Government Printing Office, 1856.

The Tale of Teleguac: José Jesús, Native Resistance, and Survival (pages 25–34)

Barbour, G. W., Redick McKee, and O. M. Wozencraft. Barbour, McKee, and Wozencraft to Luke Lea, February 17, 1851. In *Report of the Secretary of the Interior, Communicating, in Compliance with a Resolution of the Senate, a Copy of the Correspondence between the Department of the Interior and the Indian Agents and Commissioners in California*. United States Congress, 33rd Congress (1853–55), Special Session, Senate Executive Documents, Document 4, 56–59.

Burnett, Peter. "The Governor's Message." Reprint of Governor Peter Burnett's State of the State Address, January 6, 1851. *Sacramento Transcript*, January 10, 1851.

Cook, Sherburne F. "Expedition to the Interior of California, Central Valley, 1820–1840." *University of California Anthropological Records* 16, no. 6 (1960).

Gilbert, Frank T. *History of San Joaquin County, California*. Oakland: Thompson and West, 1879.

Gray, Thorne B. *Stanislaus Indian Wars: The Last of the California Northern Yokuts*. Modesto, CA: McHenry Museum Press, 1993.

Holterman, Jack. "The Revolt of Estanislao." *Indian Historian* (1969): 43–54.

Hurtado, Albert. *Indian Survival on the California Frontier*. New Haven: Yale University Press, 1988.

Kelly, William. *An Excursion to California over the Prairie, Rocky Mountains, and Great Sierra Nevada with a Stroll through the Diggings and Ranches of that Country*. 2 vols. London: Chapman and Hall, 1851.

McCarthy, Francis Florence. *The History of San Jose, California, 1797–1835*. Fresno: Academy Library Guild, 1958.

Palomares, José Francisco. "Campaign against the Rancherias of Jose de Jesus on the Tuolumne, Estanislao & Saulon on the Stanislaus, and the Moquelamos Chief Cipriano at Calaveras." In *Memoria*, 31–38. Translated by S. R. Clemence, from C. Hart Merriam Papers, vol. 1, "Relating to Work with California Indians, 1850–1974 (bulk 1898–1938)." Original manuscript at the Bancroft Library, University of California, Berkeley. Microfilm: BANC FILM 1022. Originals: BANC MSS 80/18 c.

Placer Times (Sacramento), April 28, 1849.

Rosenus, Alan. *General Vallejo and the Advent of the Americans*. Berkeley: Heyday, 1995.

Ross, John E. "Narrative of an Indian Fighter: Jacksonville, Oregon, and Related Materials." Interview by Hubert Howe Bancroft, 1878–85. Original manuscript at the Bancroft Library, University of California, Berkeley.

San Francisco Daily Alta California, May 29, 1850.

Sepulveda, Louisa. "When Santa Clara County Was Young." Interview by Mrs. Fremont [Cora] Older for a series for the *San Jose News* beginning September 30, 1925.

"A Vast Deal of Knavery": The Warnings of Daniel Walton (pages 35–39)

Cleaveland, Elisha L. "Hasting to Be Rich: A Sermon, Occasioned by the Present Excitement Respecting the Gold of California, Preached in the Cities of New Haven and Bridgeport, Jan. and Feb. 1849." New Haven, CT: J. H. Benham, 1849.

Colton, Walter. Colton private letter to John Young Mason, Secretary of the Navy, September 16, 1848. *New York Herald*, December 2, 1848.

Corpus Christi [TX] Star, January 13, 1849.

Journal des Débats (Paris), February 14, 1849.

Kalamazoo [MI] Gazette, March 29, 1849.

Memphis [TN] Eagle, May 22, 1849.

New York Herald, December 9, 1848.

Polk, James K. *State of the Union Address*. December 5, 1848.

Times (London), December 14, 1848.

Walton, Daniel. *Wonderful Facts from the Gold Regions: Also, Valuable Information Desirable to Those Who Intend Going to California*. Boston: Stacy, Richardson and Co., 1849. [The title printed on the outer wrapper is *Wonderful Facts from the Gold Regions*; the title printed on the title page is *The Book Needed for the Times, Containing the Latest Well-Authenticated Facts from the Gold Regions*.]

"The El Dorado of Their Most Sanguine Wants":
John Linville Hall, the First Journal (pages 40–46)

Hall, J. Linville. *Journal of the Hartford [CT] Union Mining and Trading Company*. Printed by J. L. Hall on board the *Henry Lee*, 1849.

New York Herald, January 24 and 29, 1849.

Upham, Samuel C. "Articles of Agreement for Perseverance Mining Company," in *Notes of a Voyage to California via Cape Horn, Together with Scenes in El Dorado, in the Years of 1849–'50*. Philadelphia: Samuel C. Upham, 1878.

"A Perfect Used Up Man": In the Diggings (pages 47–53)

Brown, Jared Comstock. Jared Comstock Brown to Charles Brown, August 11, 1851. Jared Comstock Brown Letters, 1850–57. California History Section (SMCII, box 10, folder 6), California State Library, Sacramento.

Derbec, Etienne. *A French Journalist in the California Gold Rush: The Letters of Etienne Derbec*. Georgetown, CA: Talisman Press, 1964.

"Life in California." Song in *The Gold Rush Song Book,* by Eleanora Black and Sidney Robertson. San Francisco: Colt Press, 1940.

Mulford, Prentice. *Life by Land and Sea*. New York: F. J. Needham, 1889.

Perkins, William. *Three Years in California*. Edited by Dale L. Morgan and James R. Scobie. Berkeley: University of California Press, 1964.

Shufelt, Sheldon. *A Letter from a Gold Miner, Placerville, California, October, 1850*. San Marino, CA: Friends of the Huntington Library, 1944.

Swan, John Alfred. *A Trip to the Gold Mines of California in 1848*. San Francisco: Book Club of California, 1960.

Wierzbicki, Felix. *California As It Is, and As It May Be*. San Francisco: Washington Bartlett, 1849.

Winchester, Jonas. Jonas Winchester Collection. California History Section (microfilm 2337), California State Library, Sacramento.

Wyman, Walker. *California Emigrant Letters*. New York: Bookman Associates, 1952.

*The following sources are quoted in *California Emigrant Letters*, by Walker Wyman. New York: Bookman Associates, 1952.

Letter from "Boone Emigrant," February 10, 1851. *Missouri Statesman,* April 21, 1851.

Burnett, Peter H. Burnett to Colonel A. W. Doniphan, May 1849. *Missouri Republican*, May 15, 1849.

"D.J.L." to a friend in Baltimore, Maryland. *St. Joseph [MI] Adventurer*, October 28, 1850.

Letter from W. A. George, February 12, 1850. *Missouri Republican,* April 19, 1850.

McClellan, M. T. McClellan to B. Leonard, Jackson County, Missouri, October 18, 1848. *Missouri Statesman*, April 27, 1849.

Letter from "Miffin," October 25, 1849. *St. Joseph [MI] Adventurer*, February 8, 1850.

Pratt, E. R. Pratt to his brother. *Missouri Statesman*, February 2, 1850.

Stevenson, J. D. Stevenson to his son-in-law James H. Brady in New York City, April 1849. *St. Joseph [MI] Adventurer,* July 27, 1849.

Letter from "V.J.F.," April 6, 1849. *Missouri Republican*, June 22, 1849.

Williams, A. M. Williams to his father, November 10, 1850. Originally published in the *Missouri Courier*; quoted in the *St. Joseph [MI] Adventurer*, February 21, 1851.

"With a Taint of Fraud and a Spice of Comedy": Claim Salting (pages 54–58)

Burgess, Hubert. "Anecdotes from the Mines." *Century Magazine* 42, no. 2 (June 1891): 269–70. [Some believe this story is apocryphal or exaggerated.]

Hittell, Theodore H. *History of California*. Vol. 3, 198–99. San Francisco: N. J. Stone, 1897.

Lord, Israel Shipman Pelton. *At the Extremity of Civilization: An Illinois Physician's Journey to California in 1849*. Edited by Necia Dixon Liles. Jefferson, NC: McFarland, 1885. Original journal at the Huntington Library, San Marino, CA.

The Miners' Pioneer Ten Commandments of 1849. Lithograph. Chicago: Kurz and Allison's Art Studio, 1887. Includes the note "Copyrighted in 1887 by W. P. Bennett of Gold Hill, Nevada."

Peters, Charles. *The Autobiography of Charles Peters, in 1915 the Oldest Pioneer Living in California, Who Mined in . . . the Days of '49 . . . Also Historical Happenings, Interesting Incidents and Illustrations of the Old Mining Towns in the Good Luck Era, the Placer Mining Days of the '50s*. Sacramento: LaGrave Company, c. 1915.

Schaeffer, Luther M. Quote from Schaeffer's diary entry of March 12, 1851, in *Sketches of Travels in South America, Mexico and California*. New York: J. Egbert, printer, 1860.

Swain, William. Swain to his mother, July 19, 1850. Swain Diary and Letters. Yale University Library. [Swain is extensively quoted in *The World Rushed In*, by J. S. Holliday. New York: Simon and Schuster, 1981.]

Castles in the Air: Quartz Fever (pages 59–65)

Bean, Edwin F. *Bean's History and Directory of Nevada County, California*. Nevada City: Daily Gazette, 1867.

Haskins, Charles Warren. *The Argonauts of California, Being the Reminiscences of Scenes and Incidents that Occurred in California in Early Mining Days*. New York: Fords, Howard and Hulbert, 1890.

Huntley, Henry. *California: Its Gold and Its Inhabitants*. London: T. C. Newby, 1856.

Lecouvreur, Josephine Rosana. *From East Prussia to the Golden Gate, by Frank Lecouvreur; Letters and Diary of the California Pioneer, Edited in Memory of Her Noble Husband*. Translated and compiled by Julius C. Behnke. New York: Angelina, 1906.

Richmond [IN] Palladium, February 7, 1849.

Sacramento Union, January 29, 1852; April 29, 1857; May 23, 1857; and August 15, 1857.

Sargeant, Aaron A. "Sketch of Nevada County." In *Brown and Dallison's Nevada, Grass Valley, and Rough and Ready Directory—An Almanac for 1856*, by Nathaniel P. Brown and John K. Dallison, 7–45. San Francisco: Brown and Dallison, 1856.

Woods, Daniel Bates. *Sixteen Months at the Gold Diggings*. New York: Harper and Brothers, 1851.

"Death Stared Them Full in the Face": Silas Weston and Kelly's Bar (pages 66–69)

Weston, Silas. *Life in the Mountains; or, Four Months in the Mines of California*. Providence, RI: E. P. Weston, 1854.

"That Blighting Curse": Dissipation (pages 70–76)

Borthwick, J. D. *Three Years in California*. Edinburgh: William Blackwood and Sons, 1857.

Delano, Alonzo. *Life on the Plains and among the Diggings*. Auburn, NY: Miller, Orton and Mulligan, 1856.

Fairchild, Lucius. *California Letters of Lucius Fairchild*. Madison: State Historical Society of Wisconsin, 1931.

Farnham, Eliza W. *California, In-doors and Out; or, How We Farm, Mine, and Live Generally in the Golden State*. New York: Dix, Edwards and Co., 1856.

Helper, Hinton Rowan. *The Land of Gold: Reality versus Fiction*. Baltimore: Hinton Helper, 1855.

Letts, John M. *California Illustrated: Including a Description of the Panama and Nicaragua Routes*. New York: R. T. Young, 1853.

Marryat, Frank. *Mountains and Molehills; or, Recollections of a Burnt Journal*. New York: Harper and Brothers, 1855.

Ryan, William Redmond. *Personal Adventures in Upper and Lower California, in 1848–9*. London: W. Shoberl, 1850.

Shaw, William. *Golden Dreams and Waking Realities: Being the Adventures of a Gold-Seeker in California and the Pacific Islands*. London: Smith, Elder and Co., 1851.

Soulé, Frank, John H. Gihon, and James Nisbet. *The Annals of San Francisco*. New York: D. Appleton, 1855.

Steele, James. *Old Californian Days*. Chicago: Belford-Clarke, 1889.

Wilson, Luzena Stanley. *Luzena Stanley Wilson, '49er: Memories Recalled Years Later for Her Daughter Correnah Wilson Wright*. Oakland: Eucalyptus Press, 1937.

"Plenty of Jabbering and Quarreling and Several Fights": Alfred Doten (pages 77–83)

Doten, Alfred. *The Journals of Alfred Doten, 1849–1903*. 3 vols. Reno: University of Nevada Press, 1973.

Gleason, James H. Letter to the editor. *Plymouth Rock* (Massachusetts), February 1, 1849. Alfred Doten Collection (no. NC08). University of Nevada, Reno, Collection.

"Gambling on One Card the Fruit of His Labor for the Year": Games of Chance (pages 84–89)

Borthwick, J. D. *Three Years in California*. Edinburgh: William Blackwood and Sons, 1857.

Colville, Samuel. *Colville's Marysville Directory for the Year Commencing November 1, 1855*. San Francisco: Monson and Valentine, 1855.

Hall, J. Linville. *Journal of the Hartford [CT] Union Mining and Trading Company, 1849*. Printed by J. L. Hall on board the *Henry Lee*, 1849.

Helper, Hinton Rowan. *The Land of Gold: Reality versus Fiction*. Baltimore: Hinton Helper, 1855.

Journal of the Sixth Session of the Legislature of the State of California, 1855. Sacramento: B. B. Redding, 1855.

Lord, Israel Shipman Pelton. *A Doctor's Gold Rush Journey to California*. Edited by Necia Dixon Niles. Lincoln: University of Nebraska Press, 1995. Original diary at the Huntington Library, San Marino, CA.

Pérez Rosales, Vicente. *Recuerdos del Pasado*. 1855. In *California Adventure*, by Vicente Pérez Rosales. Translated by Edwin S. Morby and Arturo Torres-Rioseco. San Francisco: Book Club of California, 1947.

Peters, William B. Peters to Miss Emily Howard, January 1, 1851. Original letter in the California History Section (SMCII, box 13, folder 12), California State Library, Sacramento.

Soulé, Frank, John H. Gihon, and James Nisbet. *The Annals of San Francisco*. New York: D. Appleton, 1855.

Thoreau, Henry David. *The Journal of Henry David Thoreau*. 1852. Vol. 1. New York: Dover, 1962.

Vicuña MacKenna, Benjamín. *Páginas de mi Diario de Viaje*. Santiago, 1862. In *We Were 49ers: Chilean Accounts of the California Gold Rush*. Translated and edited by Edwin A. Beilharz and Carlos U. López. Pasadena, CA: Ward Ritchie Press, 1976.

"There Is No Persuasion More Esteemed for Moral Conduct": Jews in the Gold Rush (pages 90–97)

Abrahamsohn, Abraham. *An Interesting Account of the Travels of Abraham Abrahamsohn to America, Especially to the Goldmines of California and Australia*. Ilmenau, Germany: Carl Friedrich Trommsdorff, 1856. Reprinted as "Interesting Accounts of the Travels of Abraham Abrahamsohn." *Western States Jewish Historical Quarterly*, April 1969.

"An Act for the Better Observance of the Sabbath." Approved April 10, 1858. *Statutes of California, Passed at the Ninth Session of the Legislature, 1858*, Chapter 171, 124–25. Sacramento: John O'Meara, State Printer, 1858.

Auerbach, Eveline Brooks. *Frontier Reminiscences of Eveline Brooks Auerbach*. Edited by Annagret S. Ogden. Berkeley: Friends of the Bancroft Library, University of California, Berkeley, 1994.

Farnham, Eliza W. *California, In-doors and Out; or, How We Farm, Mine, and Live Generally in the Gold State*. New York: Dix, Edwards and Co., 1856.

Kahn, Ava. *Jewish Voices of the California Gold Rush: A Documentary History 1849–1880*. Detroit: Wayne State University Press, 2002.

Kahn, Ava, and Marc Dollinger, eds. *California Jews*. Waltham, MA: Brandeis University Press, 2003.

Labatt, Henry. "Remarks of May 1861." *Jewish Messenger* (NY), July 17, 1861.

Levinson, Robert E. "The History of the Jews of Grass Valley, Nevada City and Vicinity." *Nevada County Historical Society Bulletin* 25, no. 3 (July 1971).

———. *The Jews in the California Gold Rush*. Berkeley: Commission for the Preservation of Pioneer Jewish Cemeteries and Landmarks of the Judah L. Magnes Museum, 1994.

Linoberg, Emanuel. Letter to the editor. *American Israelite* (Cincinnati, OH), November 13, 1857.

Rosenbaum, Fred. *Cosmopolitans: A Social and Cultural History of the Jews of the San Francisco Bay Area*. Berkeley: University of California Press, 2009.

Rosenheim, Aaron. Rosenheim to Isaac Leeser, December 28, 1852. Isaac Leeser Collection (ID: LSDCBx9ff14_17). Gershund Bennett Isaac Leeser Digitalization Project, Library of Herbert D. Katz Center for Advanced Judaic Studies, University of Pennsylvania Library.

Sarna, Jonathan. "The 'Mythical Jew' and 'The Jew Next Door' in Nineteenth-Century America." In *Anti-Semitism in American History*, edited by David A. Gerber. Urbana: University of Illinois Press, 1986.

United States Congress. House Executive Documents, 2nd Session, 51st Congress, vol. 36, 83–86 (1891).

"Everything Looks Forlorn and Wretched": Storms and Floods (pages 98–106)

Brewer, William. *Up and Down California in 1860–1864*. 1864. Berkeley: University of California Press, 1966.

Colville, Samuel. *The Sacramento Directory for the Year 1853–54*. Sacramento: Union Office, 1853. Reprinted with an introduction by Mead B. Kibbey. Sacramento: California State Library Foundation, 1997.

Hester, Sallie. "The Diary of a Pioneer Girl." Seven serialized articles. *Argonaut* (San Francisco), September 1 to October 24, 1925. [The subtitle of Hester's diary was "The Adventures of Sallie Hester, Aged Twelve, in a Trip Overland in 1849." Sallie Hester was actually fourteen at the time of her journey.]

Ingram, B. Lynn. "California Megaflood: Lessons from a Forgotten Catastrophe." *Scientific American* 308, no. 1 (December 18, 2012).

Jackson, Donald Dale. *Gold Dust*. New York: Alfred A. Knopf, 1980.

Jelinek, Lawrence James. "Property of Every Kind: Ranching and Farming during the Gold-Rush Era." *California History* 77, no. 4 (winter): 233.

Lord, Israel Shipman Pelton. *"At the Extremity of Civilization": An Illinois Physician's Journey to California in 1849*. Edited by Necia Dixon Liles. Jefferson, NC: McFarland, 1885. Original journal at the Huntington Library, San Marino, CA.

Placer Times (Sacramento), January 19, 1850.

Royce, Sarah. *A Frontier Lady*. New Haven, CT: Yale University Press, 1932.

San Francisco Daily Alta California, January 16, 1850; October 25, 1850; and December 25, 1852.

Willis, William Ladd. *History of Sacramento County, California, with Biographical Sketches of the Leading Men and Women of the County Who Have Been Identified with Its Growth and Development from the Early Days to the Present*. Los Angeles: Historic Record Company, 1913.

Winchester, Jonas. Jonas Winchester Collection. California History Section (microfilm 2237), California State Library, Sacramento.

"A Wind Turned Dark with Burning": The Plague of Fire (pages 107–113)

Borthwick, J. D. *Three Years in California*. Edinburgh: William Blackwood and Sons, 1857.

de Massey, Ernest. *A Frenchmen in the Gold Rush: The Journal of Ernest de Massey, Argonaut of 1849*. Translated by Marguerite Eyer Wilbur. San Francisco: California Historical Society, 1927.

Huntington, Collis P. Huntington to his brother Solon, November 14, 1852. Quoted in *The Great Persuader*, by David Lavender. New York: Doubleday, 1970. Original in the Collis Huntington Collection, San Marino, CA.

Huntington, Elizabeth. Huntington to her sister-in-law Phoebe Pardee, January 1, 1853. Quoted in *The Great Persuader*, by David Lavender. New York: Doubleday, 1970. Original in the Collis Huntington Collection, San Marino, CA.

Sacramento Union, November 4, 1852; November 5, 1852; December 14, 1852; and November 5, 1853.

San Francisco Daily Alta California, May 4, 1851.

Schaeffer, Luther M. Quote from Schaeffer's diary entry of June 14, 1851, in *Sketches of Travels in South America, Mexico and California*. New York: J. Egbert, printer, 1860.

"The Blue Vault of Heaven": Alonzo Delano and the Great Grass Valley Fire of 1855 (pages 114–118)

Delano, Alonzo. *Life on the Plains and among the Diggings*. Auburn, NY: Miller, Orton and Mulligan, 1856.

———.*Old Block's Sketch Book*. Sacramento: James Anthony, 1856. Reprinted with a foreword by Marguerite Wilbur. Santa Ana, CA: Fine Arts Press, 1947.

———. *Pen-Knife Sketches, or Chips of the Old Block*. Sacramento: Union Office, 1853. Reprinted with a foreword by Ezra Dane. San Francisco: Grabhorn Press, 1934.

Poingdestre, J. E., compiler. *"The Quartz Crowned Empress of the Sierras": Grass Valley and Vicinity*. Oakland: Pacific Press, 1895.

Prisk, W. F. *Nevada County Mining Review*. Grass Valley, CA: Daily Morning Union, 1895.

Thompson and West. *History of Nevada County*. Oakland: Thompson and West, 1880.

A Crumbling Kingdom: The Collapse of Sutter's Fort (pages 119–125)

Hurtado, Albert. *John Sutter: A Life on the North American Frontier*. Norman: University of Oklahoma Press, 2006.

Hutchings, James Mason. "Sutter's Account of the Gold Discovery." *Hutchings' Illustrated California Magazine* 2, no. 5 (November 1857).

McKinstry, George. McKinstry to Edward M. Kern, December 23, 1851. Quoted in *John Sutter: A Life on the North American Frontier*, by Albert Hurtado. Norman: University of Oklahoma Press, 2006. Original letter in the Fort Sacramento Papers. Huntington Library, San Marino, CA.

Prince, William Robert. William Robert Prince to Charlotte Prince, March 29, 1851. Quoted in *John Sutter: A Life on the North American Frontier,* by Albert Hurtado. Norman: University of Oklahoma Press, 2006. Original letter in the William Robert Prince Papers, 1849–51. Bancroft Library, University of California, Berkeley.

Sutter, John A. *Diary*. San Francisco: Grabhorn Press, 1932.

———. Personal reminiscences, as dictated to Hubert Howe Bancroft in 1876. Quoted in *John Sutter: A Life on the North American Frontier,* by Albert Hurtado. Norman: University of Oklahoma Press, 2006. Original manuscript in the Bancroft Library, University of California, Berkeley.

Upham, Samuel C. "Articles of Agreement for Perseverance Mining Company." In *Notes of a Voyage to California via Cape Horn, Together with Scenes in El Dorado, in the Years of 1849–'50*. Philadelphia: Samuel C. Upham, 1878.

Wiggins, William. Personal reminiscences, as dictated to Hubert Howe Bancroft in 1877. Quoted in *John Sutter: A Life on the North American Frontier,* by Albert Hurtado. Norman: University of Oklahoma Press, 2006. Original manuscript in the Bancroft Library, University of California, Berkeley.

The Devil's Chaos: Hydraulic Mining (pages 126–133)

Bancroft, Hubert Howe. *History of California*. Vol. 6, *1848–1859*. San Francisco: The History Company, 1888.

Bowles, Samuel. *Across the Continent*. Springfield, MA: Samuel Bowles and Company, 1865.

California Department of Conservation (CDOC). *California's Abandoned Mines: A Report on the Magnitude and Scope of the Issue in the State.* 2 vols. Sacramento: CDOC, 2000.

Clark, William B. *Gold Districts of California.* Sacramento: California Division of Mines and Geology, Bulletin 193 (1970).

Delta Tributary Mercury Council. *Strategic Plan for the Reduction of Mercury-Related Risk in the Sacramento River Watershed.* Sacramento: Delta Tributary Mercury Council, 2002.

Holliday, J. S. *Rush for Riches*. Berkeley: University of California Press, 1999.

Hunerlach, Michael P., James J. Rytuba, and Charles N. Alpers. "Mercury Contamination from Hydraulic Placer-Gold Mining in the Dutch Flat Mining District, California." In *United States Geological Survey Water-Resources Investigation, Report 99-4018B* (1999), 179–89.

Kelley, Robert. *Gold v. Grain*. Glendale, CA: Arthur H. Clark, 1959.

Marysville [CA] Appeal, January 8, 1884.

San Francisco Daily Evening Bulletin, January 8, 1884.

Sierra Fund. *Mining's Toxic Legacy*. Nevada City, CA: Sierra Fund, 2008.

St. Clair, David J. "The Gold Rush and the Beginnings of California Industry." In *A Golden State: Mining and Economic Development in Gold Rush California*, edited by James Rawls and Richard Orsi, 185–208. Berkeley: University of California Press, 1998.

United States Congress. House Executive Documents, 2nd Session, 51st Congress, vol. 36, 83–86 (1891).

Vischer, Edward. "A Trip to the Mining Regions in the Spring of 1859." Part 2. *California Historical Society Quarterly* 2 (December 1932): 332.

von Geldern, Otto. *An Analysis of the Problem of the Proposed Rehabilitation of Hydraulic Mining in California*, p. 4. Sutter County: January 3, 1928.

Woodruff v. North Bloomfield Gravel Mining Co. and Others. *Federal Reporter* 18 (1884), cited as 9 Sawyer 441 or 18 F 753.

"As Huge, to Me, as an Elephant": Grizzly Bears (pages 134–140)

Bell, Horace. *Reminiscences of a Ranger; or, Early Times in Southern California.* Los Angeles: Yarnell, Caystile and Mathes, 1881.

Day, Mrs. F. H. "Sketches of the Early Settlers of California: George C. Yount," *The Hesperian* 2, no. 1 (March 1859).

Frémont, Jessie Benton. *Far-West Sketches*. Boston: D. Lothrop, 1890.

"The Grizzly Bear of California." *Harper's New Monthly Magazine* 15, no. 90 (November 1857).

Gudde, Edwin. *California Place Names: A Geographical Dictionary*. Berkeley: University of California Press, 1949, 1960, 1969.

Haun, Catherine. "A Woman's Trip Across the Plains, 1849." Original diary at the Huntington Library, San Marino, CA. Reprinted in part in *Women's Diaries of the Westward Journey*, by Lillian Schlissel. New York: Schocken Books, 1992.

Hines, Joseph Wilkinson. *Touching Incidents in the Life and Labors of a Pioneer on the Pacific Coast since 1853*. San Jose: Eaton and Company, 1911.

Hutchings, James Mason. "The Grizzly Bear." *Hutchings' Illustrated California Magazine* 1, no. 3 (September 1856).

Möllhausen, Balduin. *Diary of a Journey from the Mississippi to the Coasts of the Pacific.* London: Longman, Brown, Green and Roberts, 1858.

Muir, John. *Our National Parks*. Boston: Houghton-Mifflin, 1901.

Perkins, William. *Three Years in California*. Edited by Dale L. Morgan and James R. Scobie. Berkeley: University of California Press, 1964.

Snyder, Susan, ed. *Bear in Mind: The California Grizzly*. Berkeley: Bancroft Library and Heyday, 2003.

Taylor, Bayard. *Eldorado: Adventures in the Path of Empire*. New York: G. P. Putnam, 1850.

Whipple-Haslam, Lee. *Early Days in California; Scenes and Events of the '50s as I Remember Them*. Jamestown, CA: 1925.

"A Very Normal Childhood": Children and Families (pages 141–145)

Barry, Theodore A., and Patten, Benjamin A. *Men and Memories of San Francisco, in the "Spring of '50."* San Francisco: A. L. Bancroft, 1856.

Gunn, Elizabeth Le Breton. Gunn to her family back East, August 24, 1851. In *Records of a California Family: Journals and Letters of Lewis C. Gunn and Elizabeth Le Breton Gunn*, edited by Anna Lee Marston. San Diego: 1928.

Huntley, Henry. *California: Its Gold and Its Inhabitants*. London: T. C. Newby, 1856.

Marston, Anna Lee [Gunn], ed. *Records of a California Family: Journals and Letters of Lewis C. Gunn and Elizabeth Le Breton Gunn*. San Diego: 1928. [Three hundred copies of this book were printed for Anna Lee Gunn Marston by Johnck and Seeger in San Francisco in November 1928. Marston was the daughter of Lewis Carstairs Gunn and Elizabeth Le Breton Stickney Gunn.]

Mulford, Prentice. *Life by Land and Sea*. New York: F. J. Needham, 1889.

Shaw, William. *Golden Dreams and Waking Realities: Being the Adventures of a Gold-Seeker in California and the Pacific Islands*. London: Smith, Elder and Co., 1851.

White, William Francis. *A Picture of Pioneer Times in California, Illustrated with Anecdotes and Stories Taken from Real Life. By William Grey* [pseudonym]. San Francisco: Printed by W. M. Hinton and Company, 1881.

"The Years Have Been Full of Hardships": Luzena Stanley Wilson (pages 146–154)

Levy, JoAnn. *They Saw the Elephant: Women in the California Gold Rush*. Norman: University of Oklahoma Press, 1992.

Wilson, Luzena Stanley. *Luzena Stanley Wilson, '49er: Memories Recalled Years Later for Her Daughter Correnah Wilson Wright*. Oakland: Eucalyptus Press, 1937.

Shadow and Light: Mifflin Wistar Gibbs and Defiance of Discrimination (pages 155–162)

"An Act to Restrict and Prevent the Immigration to and Residence in This State of Negroes and Mulattoes." Assembly Bill 339, California State Legislature, introduced on March 19, 1858, by Assemblyman J. B. Warfield of Nevada County. *Journal of the Ninth Session of the Assembly of the State of California (January 4–April 26, 1858)*, 408. Sacramento: John O'Meara, State Printer, 1858.

Dred Scott v. John F. A. Sanford. Judgment in the U.S. Supreme Court Case, March 6, 1857; Case Files 1792–1995; Record Group 267; Records of the Supreme Court of the United States; National Archives.

Gibbs, Mifflin Wistar. *Shadow and Light: An Autobiography.* Washington, D.C.: 1902. Reprint, Lincoln: University of Nebraska Press, 1995.

Grodin, Joseph R. "The California Supreme Court and State Constitutional Rights: The Early Years." *Hastings Constitutional Law Quarterly* 31, no. 2 (2004): 141–62.

Lapp, Rudolph M. *Blacks in Gold Rush California*. New Haven: Yale University Press, 1977.

Matter of Archy, 9 Cal. 147 (1858). Decision of the California State Supreme Court.

Pilton, James William. *Negro Settlement in British Columbia, 1858–1871.* MA thesis, University of British Columbia, 1951.

Proceedings of the California State Convention of Colored Citizens. San Francisco: Printed at the Office of "The Elevator," corner of Sansome and Jackson Streets, 1865. Reprint, San Francisco: R&E Research Associates, 1969.

Proceedings of the First State Convention of the Colored Citizens of the State of California. Sacramento: Democratic State Journal Print, 1855. Reprint, San Francisco: R&E Research Associates, 1969.

Proceedings of the Second Annual Convention of the Colored Citizens of the State of California. San Francisco: J. H. Udell and W. Randall, Printers, 1856. Reprint, San Francisco: R&E Research Associates, 1969.

Sacramento Daily Union, November 22, 1855.

Sacramento Daily Union, March 20, 1858. Copy of Assembly Bill 339, "An Act to Restrict and Prevent the Immigration to and Residence in This State of Negroes and Mulattoes."

Sacramento Daily Union, April 16, 1858.

San Francisco Daily Alta California, February 14, 1858.

San Francisco Daily Evening Bulletin, April 21, 1858, and May 13, 1858.

A Vast, Glowing Empty Page: Reinvention and the Veritable Squibob
(pages 163–169)

Buck, Franklin A. *A Yankee Trader in the Gold Rush: The Letters of Franklin A. Buck.* Compiled by Katherine A. White. Boston: Houghton Mifflin Company, 1930.

Derby, George Horatio. *Phoenixiana; or, Sketches and Burlesques.* New York: D. Appleton, 1856. Reprint, 1902. [Derby's pseudonyms were John Phoenix and the Veritable Squibob.]

———. *The Squibob Papers*. New York: Carleton, 1865.

Farnham, Eliza W. *California, In-doors and Out; or, How We Farm, Mine, and Live Generally in the Golden State*. New York: Dix, Edwards and Co., 1856.

Fisher, Walter M. *The Californians*. London: MacMillan, 1876.

Fitzgerald, Oscar Penn. *California Sketches: New Series*. Nashville: Southern Methodist Publishing, 1881.

Holliday, J. S. *The World Rushed In: The California Gold Rush Experience*. New York: Simon and Schuster, 1981.

Jensen, Eric Frederick. "George Horatio Derby's 'Musical Review Extraordinary,' or, Félician David in the New World." *American Music* 8, no. 3 (Autumn 1990): 351–58. Published by the University of Illinois Press.

Pierce, Hiram Dwight. *A Forty-Niner Speaks: A Chronological Record of a New Yorker and His Adventures in Various Mining Localities in California, His Return Trip across Nicaragua, including Several Descriptions of the Changes in San Francisco and Other Mining Centers from March 1849 to January 1851*. Oakland: Keystone-Inglett Printing Company, 1930.

Report of the Secretary of War Communicating Information in Relation to the Geology and Topography of California. United States Senate, 31st Congress, Ex. Doc. No. 47 (April 3, 1850).

Stewart, George R. *John Phoenix, Esq., The Veritable Squibob: A Life of Captain George H. Derby, U.S.A.* New York: Henry Holt and Co., 1937.

Taylor, William. *California Life Illustrated, by William Taylor, of the California Conference*. New York: Published for the author by Carlton and Porter, 1858.

"Emperor of These United States": Joshua Norton (pages 170–177)

Clemens, Samuel [Mark Twain].Clemens to Orion and Mollie Clemens, October 19 and 20, 1865 (UCCL 00092). In *Mark Twain's Letters, 1853–1866*. Edited by Edgar Marquess Branch, Michael B. Frank, Kenneth M. Sanderson, Harriet Elinor Smith, Lin Salamo, and Richard Bucci. Mark Twain Project Online. Berkeley: University of California Press, 1988, 2007.

———.Clemens to William Dean Howells, September 3, 1880 (UCCL 01829). In *Mark Twain's Letters, 1876–1880*. Edited by Michael B. Frank and Harriet Elinor Smith. Mark Twain Project Online. Berkeley: University of California Press, 2002, 2007.

Drury, William. *Norton I, Emperor of the United States*. New York: Dodd, Mead and Company, 1986.

Fisher, Walter M. *The Californians*. London: MacMillan, 1876.

Fitzgerald, Oscar Penn. *California Sketches: New Series*. Nashville: Southern Methodist Publishing, 1881.

Kramer, William M. *Emperor Norton of San Francisco*. Santa Monica: North B. Stern, 1974.

Lane, Allen Stanley. *Emperor Norton, Mad Monarch of Montgomery Street*. Caldwell, ID: Claxton Printers, 1939.

Moylan, Peter. "Emperor Norton." *Encyclopedia of San Francisco*, www.sfhistoryencyclopedia.com.

Sacramento Daily Union, January 10, 1880.

San Francisco Alta California, January 22, 1867.

San Francisco Bulletin. "An Emperor Among Us." September 17, 1859. Reprinted in the *Red Bluff [CA] Beacon*, September 28, 1859.

Taylor, William. *California Life Illustrated, by William Taylor, of the California Conference*. New York: Published for the author by Carlton and Porter, 1858.

"Very Little Law of Any Kind": Lawyers and Judges (pages 178–182)

Bancroft, Hubert Howe. "California Inter Pocula." *The Works of Hubert Howe Bancroft, Volume 35*. San Francisco: The History Company, 1888.

Borthwick, J. D. *Three Years in California*. Edinburgh: William Blackwood and Sons, 1857.

Crosby, Elisha Oscar. *Memoirs of Elisha Oscar Crosby: Reminiscences of California and Guatemala from 1849 to 1864*. San Marino, CA: Huntington Library, 1945.

Helper, Hinton Rowan. *The Land of Gold: Reality versus Fiction*. Baltimore: Hinton Helper, 1855.

Shuck, Oscar Tully. *Bench and Bar in California: History, Anecdotes, Reminiscences*. San Francisco: Occident, 1889.

Soulé, Frank, John H. Gihon, and James Nisbet. *The Annals of San Francisco*. New York: D. Appleton, 1855.

Warren, Louis A. *Lincoln's Youth: Indiana Years, Seven to Twenty-One, 1816–1830*. Indianapolis: Indiana Historical Society, 1991.

White, William Francis. *A Picture of Pioneer Times in California, Illustrated with Anecdotes and Stories Taken from Real Life. By William Grey* [pseudonym]. San Francisco: W. M. Hinton and Company, 1881.

"Enjoy It with Luxurious Zest": James Mason Hutchings (pages 183–190)

Hutchings, James Mason. "Autobiography and Reminiscence of James Mason Hutchings, San Francisco, 1901." *Autobiographies and Reminiscences of California Pioneers* 2: 5–11. Compiled by the Historical Committee of the Society of California Pioneers.

———. *In the Heart of the Sierras*. Oakland and Yosemite Valley: Pacific Press Publishing and Old Cabin, Yosemite Valley, 1888.

———. "Off to the Mountains." *Hutchings' Illustrated California Magazine* 4, no. 4 (October 1859).

———. *Souvenir of California: Yo Semite Valley and the Big Trees*. San Francisco: J. M. Hutchings, 1896.

Hutchings, James Mason, and Harrison Eastman, illustrator. *The Miners' Ten Commandments.* San Francisco: Hutchings and Sun Print, 1853.

Kruska, Dennis. *James Mason Hutchings of Yo Semite*. San Francisco: Book Club of California, 2009.

Mariposa [CA] Gazette, August 9, 1855.

New Orleans Daily Picayune, December 19, 1849.

Letter of the Law: Letter Sheets and the Ten Commandments (pages 191–195)

Butler, W. C., engraver. *Commandments to California Wives.* San Francisco: Hutchings and Mercantile Job Print, 1855.

Cleveland Plain Dealer, March 27, 1849.

Hutchings, James Mason, and Harrison Eastman, illustrator. *The Miners' Ten Commandments.* San Francisco: Hutchings and Sun Print, 1853.

The Miners' Pioneer Ten Commandments of 1849. Lithograph. Chicago: Kurz and Allison's Art Studio, 1887. Includes the note "Copyrighted in 1887 by W. P. Bennett of Gold Hill, Nevada."

Placerville [CA] Herald, June 1853.

"To the Land of Gold and Wickedness": Lorena Lenity Hays Bowmer (pages 196–202)

Bowmer, Lorena Lenity Hays. *To the Land of Gold and Wickedness: The 1848–54 Diary of Lorena L. Hays*. Edited by Jeanne Hamilton Watson. St. Louis: Patrice Press, 1988.

Brown, William A. Brown to his father, July 25, 1850. Huntington Library, San Marino, CA.

Crackbon, Joseph. Diary entry for April 2, 1850, from *Narrative of a Voyage from New York to California via Chagres, Gorgona & Panama: Journey Across the Isthmus, etc.; Residence in Panama.* Copied from Crackbon's original diary. California History Section (BC88), California State Library, Sacramento. Quoted in *The World Rushed In*, by J. S. Holliday, 354–55. New York: Simon and Schuster, 1981.

Golden Era (San Francisco), April 8, 1855, and October 7, 1855.

Levy, JoAnn. *They Saw the Elephant: Women in the California Gold Rush.* Norman: University of Oklahoma Press, 1992.

Prince, William. Prince to his wife, October 21, 1849. Bancroft Library, University of California, Berkeley.

Shirley, Dame [Louise Amelia Knapp Smith Clappe]. Dame Shirley to her sister Molly in Massachusetts, September 20, 1851. *The Pioneer,* April 1854. Reprinted in *The Shirley Letters: From the California Mines, 1850–1852*. Berkeley: Heyday, 1998, 2001.

Wilson, Luzena Stanley. *Luzena Stanley Wilson, '49er: Memories Recalled Years Later for Her Daughter Correnah Wilson Wright.* Oakland: Eucalyptus Press, 1937.

"The Theatre of Unrest": Eliza Farnham (pages 203–211)

Farnham, Eliza W. *California, In-doors and Out; or, How We Farm, Mine, and Live Generally in the Golden State.* New York: Dix, Edwards and Co., 1856.

———. *Circular of the California Association of American Women, February 20, 1849.* New York: Nesbitt, 1849. Original at the California Historical Society, Vault B-004, ID# 00020092.

Kirby, Ora. "Recollections." *Santa Cruz [CA] Daily Surf*, June 16, 1893.

Levy, JoAnn. *Unsettling the West: Eliza Farnham and Georgiana Bruce Kirby in Frontier California*. Berkeley: Heyday, 2004.

New York Herald, April 12, 1849.

New York Tribune, February 14, 1849.

New York Tribune, February 14, 1850. Letter from a passenger aboard the *Angelique*, written December 29, 1849.

Notable American Women, 1607–1950: A Biographical Dictionary. 3 vols. New York: Belknap, 1971.

Placer Times (Sacramento), July 28, 1849.

San Francisco Californian, September 16, 1848.

San Francisco Daily Alta California, May 24, 1849; October 1, 1849; December 14, 1849; January 18, 1850; May 16, 1850; and November 7, 1851.

Sedgwick, Catharine. Mary E. Dewey, editor, to Catharine Sedgwick, February 5, 1849. *Life and Letters of Catharine M. Sedgwick.* New York: Harper and Bros., 1872.

Wierzbicki, Felix. *California As It Is, and As It May Be*. San Francisco: Washington Bartlett, 1849.

Draped in a Butterfly Robe: The Mariposa County Courthouse (pages 212–217)

Bancroft, Hubert Howe. "Essays and Miscellany." *The Works of Hubert Howe Bancroft, Volume 38*. San Francisco: The History Company, 1890.

Chamberlain, Newell. *The Call of Gold*. Mariposa, CA: Gazette Press, 1936.

"Convulsive Throes and Conflicts of Passion": Dr. John Morse, the First Historian (pages 218–226)

Dwinelle, John W. "Eulogy of Past Grand Master John F. Morse." *Proceedings of the Twenty-Second Annual Communication of the R.W. Grand Lodge of the Independent Order of Oddfellows of the State of California, Volume 5*. San Francisco: Jos. Winterburn, 1874.

Moerenhout, Jacques Antoine. *The Inside Story of the Gold Rush*. San Francisco: California Historical Society, 1935.

Morse, John Frederick. "History of Sacramento." In *The Sacramento Directory for the Year 1853–54*, by Samuel Colville. Sacramento: Union Office, 1853. Reprinted with an introduction by Mead B. Kibbey. Sacramento: California State Library Foundation, 1997.

Sacramento Union, January 1, 1875.

Shaw, Pringle. *Ramblings in California*. Toronto, Canada: J. Bain, c. 1857.

Cheesquatalawny: John Rollin Ridge (pages 227–233)

Grass Valley [CA] National. "John Rollin Ridge Obituary." October 8, 1867.

McCulley, Johnston. "The Curse of Capistrano." *All-Story Weekly* 2 (August 1919).

Parins, James W. *John Rollin Ridge: His Life and Works*. Lincoln: University of Nebraska Press, 1991. Revised 2004.

Ridge, John Rollin. *The Life and Adventures of Joaquín Murieta, the Celebrated California Bandit*. San Francisco: W. B. Cooke, 1854.

I Am Joaquín!: The Legacy of Joaquín Murieta, the Celebrated California Bandit (pages 234–242)

"An Act to Authorize the Raising of a Company of Rangers." Approved May 17, 1853. *Journal of the Fourth Session of the Legislature of the State of California*. San Francisco: George Kerr, State Printer, 1853.

Brands, H. W. *The Age of Gold*. New York: Doubleday, 2002.

Latta, Frank. *Joaquín Murietta and His Horse Gangs*. Santa Cruz, CA: Bear State Books, 1980.

New York Tribune. "California." June 14, 1853.

Ridge, John Rollin. Ridge to his mother, October 4, 1850. *Fort Smith [AR] Herald*, January 24 and 31, 1851.

———. *The Life and Adventures of Joaquín Murieta, the Celebrated California Bandit*. San Francisco: W. B. Cooke, 1854.

Vallejo, Mariano Guadalupe. Translated quote from *Recuerdos Históricos y Personales Tocante a la Alta California* [5 vols; 1875; original in the Bancroft Library, University of California, Berkeley], in *The Course of Empire: Firsthand Accounts of California in the Days of the Gold Rush of '49*, edited by Valeska Bari, 58–59. New York: Coward-McCann, 1931.

Cock-Eye and Snowshoe:
John Calhoun Johnson and Jon Torsteinson-Rue, Trailblazers (pages 243–249)

DeQuille, Dan [William Wright]. "Snow-Shoe Thompson." *Overland Monthly* 8, no. 46 (October 1886): 419–35.

Historical Souvenir of El Dorado County. Oakland: Prolo Sioli, 1883.

"How We Get Our Letters." Letter sheet published by Leland and McCombe, San Francisco, 1854. Original in the California History Section (Picture Collection: Prints: Lettersheets: Scene at San Francisco Post Office, 1999-0029), California State Library, Sacramento.

Hutchings, James Mason. "Sierra Dog Express." *Hutchings' Illustrated California Magazine* 5, no. 56 (February 1861).

James, William F., and George H. McCurry. *History of San José, California*. San José: A. H. Cawston, 1933.

Marysville [CA] Appeal, February 28, 1867.

Newmark, Harris. *Sixty Years in Southern California, 1853–1913*. New York: Knickerbocker Press, 1926.

Placerville [CA] Herald, July 9, 1853.

Sacramento Union, April 7, 1869; March 10, 1869; and March 26, 1869.

Stillman, Richard Thomas. *Spreading the Word: A History of Information in the California Gold Rush*. Lincoln: University of Nebraska, 2006.

Dueling Designations: Lake Bigler versus Lake Tahoe (pages 250–254)

Assembly Bill 437. *Journal of the House of Assembly of the State of California at the Twelfth Session of the Legislature, 1861*. Sacramento: C. T. Botts, State Printer, 1861.

Baker, George Holbrook. *Map of the Mining Regions of California*. San Francisco: Barber and Baker, 1855. Many sources list "Maheon Lake," but it is actually "Mahlon Lake."

King, Thomas Starr. *Christianity and Humanity: A Series of Lectures*. Boston: James R. Osgood, 1878.

Sacramento Union, May 28, 1863; July 27, 1863; and February 5, 1870.

Sacramento Union, May 30, 1863. Quoting the *Marysville [CA] Appeal*.

Sacramento Union, June 1, 1863. Quoting the *Nevada Transcript*.

Sacramento Union, August 24, 1886. Quoting "'Tahoe' as She Is Spoke," *Truckee [CA] Republican*.

Senate Bill 1265. *[California] Senate Final History, Fifty-Sixth Session, 1945*. Sacramento: State Printing Office, 1945.

Shaw, David Augustus. *Eldorado; or, California as Seen by a Pioneer, 1850–1900*. Los Angeles: B. R. Baumgardt, 1900.

Statutes of California, 1869–1870, Chapter 58. Sacramento: D. W. Gelwicks, 1870.

Virginia City [NV] Territorial Enterprise, September 4, 1863.

"A Great Deal of Hard Times": The Diary of Joseph Pike (pages 255–259)

Carson, James H. *Early Recollections of the Mines, and a Description of the Great Tulare Valley*. Reprint, Tarrytown, NY: W. Abbatt, 1931. Originally printed as a supplement in the *San Joaquin Republican* (Stockton, CA) in 1852.

Pike, Joseph. *Diary of a Forty-Niner, Being the Diary of Joseph Pike of Half Day, Illinois, April 15, 1850, to December 29, 1851*. Original manuscript and typescript in the California History Section, California State Library, Sacramento.

PORTS (Parks Online Resources for Teachers and Students). "Gold Rush Prices Worksheet." California Department of Parks and Recreation, Sacramento. http://ports.parks.ca.gov/pages/22922/files/worksheet-goldrush-prices.pdf.

Guilty of Dust and Sin: Drinking, Dining, and Desire (pages 260–269)

Anonymous. "Letter from Anonymous to Lizzie," October 15, 1853. Letters Collection of the Huntington Library, San Marino, CA.

Booth, Edmund. *Edmund Booth, Forty-Niner: The Life Story of a Deaf Pioneer, including Portions of His Autobiographical Notes and Gold Rush Diary, and Selections from Family Letters and Reminiscences*. Stockton, CA: San Joaquin Pioneer and Historical Society, 1953.

Borthwick, J. D. *Three Years in California*. Edinburgh: William Blackwood and Sons, 1857.

Farnham, Eliza W. *California, In-doors and Out; or, How We Farm, Mine, and Live Generally in the Golden State*. New York: Dix, Edwards and Co., 1856.

Fitzgerald, Oscar Penn. *California Sketches: New Series*. Nashville: Southern Methodist Publishing, 1881.

Helper, Hinton Rowan. *The Land of Gold: Reality versus Fiction*. Baltimore: Hinton Helper, 1855.

Johnson, Theodore. *Sights in the Gold Region, and Scenes by the Way*. New York: Baker and Scribner, 1849.

Kip, Leonard. *California Sketches, with Recollections of the Gold Mines*. Los Angeles: N. A. Kovach, 1946. Originally published as a pamphlet in 1850.

Letts, John M. *California Illustrated: Including a Description of the Panama and Nicaragua Routes.* New York: R. T. Young, 1853.

Lord, Israel Shipman Pelton. *A Doctor's Gold Rush Journey to California.* Edited by Necia Dixon Niles. Lincoln: University of Nebraska Press, 1995. Original diary at the Huntington Library, San Marino, CA.

Marryat, Frank. *Mountains and Molehills; or, Recollections of a Burnt Journal.* New York: Harper and Brothers, 1855.

Packer, Henry. Packer to Mary Elizabeth Judkins, May 20, 1851. Letters Collection of Henry B. Packer, Bancroft Library, University of California, Berkeley.

Perkins, William. *Three Years in California.* Edited by Dale L. Morgan and James R. Scobie. Berkeley: University of California Press, 1964.

Rohrbough, Malcolm. *Days of Gold: The California Gold Rush and the American Nation.* Berkeley: University of California Press, 1997.

Shaw, William. *Golden Dreams and Waking Realities: Being the Adventures of a Gold-Seeker in California and the Pacific Islands.* London: Smith, Elder and Co., 1851.

Shirley, Dame [Louise Amelia Knapp Smith Clappe]. Dame Shirley to her sister Molly in Massachusetts, September 22, 1851. *The Pioneer,* January 1854–December 1855. Reprinted in *The Shirley Letters: From the California Mines, 1850–1852.* Berkeley: Heyday, 1998, 2001.

Soulé, Frank, John H. Gihon, and James Nisbet. *The Annals of San Francisco.* New York: D. Appleton, 1855.

Stevens, Simon. Stevens to his cousin in Rockland, Maine, June 5, 1853. Original in the California History Section (SMC II, box 12, folder 1), California State Library, Sacramento.

Taylor, Bayard. *Eldorado: Adventures in the Path of Empire.* New York: G. P. Putnam, 1850.

Thompson and West. *History of Nevada County.* Oakland: Thompson and West, 1880.

Players and Painted Stages: Curiosities and Amusements (pages 270–277)

Ayers, James J. *Gold and Sunshine: Reminiscences of Early California.* Boston: R. G. Badger, 1922.

Borthwick, J. D. *Three Years in California.* Edinburgh: William Blackwood and Sons, 1857.

Gagey, Edmond M. *The San Francisco Stage: A History.* New York: Columbia University Press, 1950.

Hill, Jasper Smith. *The Letters of a Young Miner, 1849–1852.* San Francisco: John Howell Books, 1964.

Kurutz, Gary F. "Popular Culture on the Golden Shore." In *Rooted in Barbarous Soil: People, Culture, and Community in Gold Rush California*, edited by Kevin Starr and Richard J. Orsi, 280–315. Berkeley: University of California Press, 2000.

Sacramento Union, September 20, 1854.

San Francisco Daily Alta California, August 14, 1851; June 6, 1852; June 30, 1852; July 15, 1852; and December 18, 1854.

"So Wonderful, So Dangerous, So Magnificent a Chaos": The 1849 California Constitution (pages 278–285)

Bancroft, Hubert Howe. *History of California*. Vol. 23. San Francisco: The History Company, 1888.

Browne, J. Ross. *Report of the Debates in the Convention of California, on the Formation of the State Constitution, in September and October, 1849*. Washington, D.C.: Printed by J. T. Towers, 1850.

Colton, Walter. *Three Years in California*. New York: A. S. Barnes, 1850.

Ellison, William Henry. *A Self-Governing Dominion: California, 1849–1860*. Berkeley: University of California Press, 1950.

"Memorable American Duels." *Frank Leslie's Popular Monthly* 3, no. 3 (March 1877): 267.

Milner, John T. "John T. Milner's Trip to California, Letters to the Family." *Alabama Historical Quarterly* 20, no. 3 (1958): 523–56.

New York Tribune, October 22, 1849. Article on attitudes of California gold seekers. Quoted in many sources, including *The Age of Gold*, by H. W. Brands (New York: Doubleday, 2002) and *Frémont: Pathmarker of the West*, by Allan Nevins (New York: D. Appleton, 1939).

San Francisco Alta California, May 3, 1849, and May 24, 1849.

Taylor, Bayard. *Eldorado: Adventures in the Path of Empire*. New York: G. P. Putnam, 1850.

The Legislature of a Thousand Drinks: Thomas Jefferson Green and the First California Legislature (pages 286–290)

Bancroft, Hubert Howe. *History of California, 1848–1859*. San Francisco: The History Company, 1888, 6:311.

Green, Stephen. "T. J. Green and 'The Legislature of a Thousand Drinks.'" *California State Library Foundation Bulletin*, no. 85 (2006). Sacramento: California State Library Foundation.

Hall, Frederic. *The History of San José and Surroundings*. San Francisco: A. L. Bancroft, 1871. See also a review of Hall's book in *Pacific Rural Press* 1, no. 3 (January 21, 1871).

Hittell, Theodore H. *History of California*, 3:198–99. San Francisco: N. J. Stone, 1897.

"An Act for the Better Regulation of the Mines and the Government of Foreign Miners": The Foreign Miners' Tax and the French Revolution (pages 291–298)

"An Act for the Better Regulation of the Mines and the Government of Foreign Miners." *Statutes of California Passed at the First Session of the Legislature, December 15, 1849–April 22, 1850, at the Pueblo de San José,* Chapter 97, 221. San José: J. Winchester, State Printer, 1850.

"An Act to Establish a Public Resources Code, Thereby Consolidating the Law Relating to Natural Resources, the Conservation, Utilization and Supervision Thereof, and Matters Incidental Thereto, and to Repeal Certain Acts and Parts of Acts Specified Herein." *Statutes of California, Fifty-Third Session of the Legislature, 1939*, Chapter 93 (April 26, 1939). Sacramento: State Printing Office, 1939.

"An Act to Provide for the Protection of Foreigners and to Define Their Liabilities and Privileges." *Statutes of California Passed at the Fourth Session of the Legislature*, Chapter 44 (March 30, 1853). San Francisco: George Kerr, State Printer, 1853.

California State Controller's Reports, 1850–1870. Figures found in *Chinese Immigration*, by Mary Roberts Coolidge, 37 (New York: Henry Holt and Co., 1909) and *A History of Two Chinatowns in Grass Valley and Nevada City*, by Thomas Arthur Deeble (MA thesis, San Francisco State University, 1972). These publications feature statistics related to the Foreign Miners' Tax from 1850 to 1870. Originally the tax collector was apportioned part of the total collections as payment, and in later years other government agencies received a portion of the total collections as well. From 1852 to 1854, the tax collector was allowed 10 percent of the total. From 1854 to 1861, the tax collector was allowed 15 percent, the sheriff 3 percent, and the recorder 3 percent. From 1861 to 1869, the tax collector was allotted 20 percent of total collections.

Dèlépine, Antoine Alphonse. Dèlépine to his father, August 20, 1850. Dèléphine Papers, California State Library, Sacramento. See also *To My Children: A Simple Narrative of My Travels*, by Antoine Alphonse Dèlépine. Sonora, CA: Banner Printing Company for Harold Mojonnier, 1963.

Gerstäcker, Friedrich. *Scenes of Life in California.* Translated from the French by George Cosgrave. San Francisco: J. Howell, 1942. Originally published in Germany in 1856 as *Californische Skizzen*.

Harris, Benjamin Butler. *The Gila Trail: The Texas Argonauts and the California Gold Rush.* Norman: University of Oklahoma Press, 1960.

Heckendorn, J., and W. A. Wilson. *Miners and Business Men's Directory for the Year Commencing January 1, 1856, Embracing a General History of the Citizens of Tuolumne . . . Together with the Mining Laws of Each District, a Description of the Different Camps, and Other Interesting Statistical Matter.* Columbia, CA: Clipper Office, 1856.

"Report of Mr. Green on Mines and Foreign Miners, in Senate, March 15, 1850." *Journal of the Senate of the State of California at Their First Session, Begun and Held at the Puebla de San José on the Fifteenth Day of December 1849,* 493–97. San José: J. Winchester, State Printer, 1850.

Rohrbough, Malcolm. *Days of Gold: The California Gold Rush and the American Nation.* Berkeley: University of California Press, 1997.

Woods, Daniel Bates. *Sixteen Months at the Gold Diggings*. New York: Harper and Brothers, 1851.

"We Are Not the Degraded Race You Would Make Us": Hab Wa, Long Achick, Noman Asing, and Governor John Bigler (pages 299–306)

Asing, Noman [Yuen Sheng]. "To His Excellency Governor Bigler." *San Francisco Daily Alta California*, May 5, 1852. [Many sources refer to "Norman Asing" but the letter in the *Daily Alta California* is signed "Noman Asing."]

Hab Wa and Long Achick. "The Chinese in California: Letter of the Chinamen to His Excellency Gov. Bigler." *Living Age* (San Francisco) 34 (April 29, 1852).

McLain, Charles J., Jr. "The Chinese Struggle for Civil Rights in 19th Century America: The First Phase, 1850–1870." *California Law Review* 72, no. 4, article 7 (July 1984).

Pfaelzer, Jean. *Driven Out: The Forgotten War against Chinese Americans.* Berkeley: University of California Press, 2008.

"Report of the Joint Select Committee Relative to the Chinese Population of the State of California." *Journals of the [California] Senate and Assembly*, Appendix, vol. 3, p. 7. Sacramento: State Printing Office, 1862.

San Francisco Daily Alta California, November 23, 1858. Quoting the article "The Wrongs of Chinamen," from an 1857 issue of the *Placerville [CA] American.*

"The Theatre of Scenes which We Hope Will Never Be Repeated": Vigilantes (pages 307–312)

"An Act Concerning Crimes and Punishments." *Statutes of California Passed at the First Session of the Legislature, December 15, 1849–April 22, 1850, at the Pueblo de San José*, Chapter 99, 247 (1850). San José: J. Winchester, State Printer, 1850.

Amendment to "An Act Concerning Crimes and Punishments." *Statutes of California Passed at the Second Session of the Legislature, January 6, 1851–May 1, 1851, at the City of San José*, Chapter 95, 406 (1851). San José: Eugene Casselly, State Printer, 1851.

"An Act for the Better Regulation of the Mines and the Government of Foreign Miners." *Statutes of California Passed at the First Session of the Legislature, December 15, 1849–April 22, 1850, at the Pueblo de San José*, Chapter 97, 221–23 (1850). San José: J. Winchester, State Printer, 1850.

Bancroft, Hubert Howe. *Popular Tribunals*. 2 vols. San Francisco: The History Company, 1887.

Evans, Albert S. *À la California: Sketch of Life in the Golden State*. San Francisco: A. L. Bancroft and Company, 1873.

Gonzales-Day, Ken. *Lynching in the West: 1850–1935*. Durham, NC: Duke University Press, 2006.

Perkins, William. *Three Years in California*. Edited by Dale L. Morgan and James R. Scobie. Berkeley: University of California Press, 1964.

"Report of Mr. Green on Mines and Foreign Miners, in Senate, March 15, 1850." *Journal of the Senate of the State of California at Their First Session, Begun and Held at the Puebla de San José on the Fifteenth Day of December 1849*, 493–97. San José: J. Winchester, State Printer, 1850.

Sacramento News-Letter, fortnight ending August 30, 1851.

Sacramento Union, August 23, 1851, and December 26, 1885.

San Francisco Evening Picayune, August 1850. Quoted in *Popular Tribunals*, by Hubert Howe Bancroft. 2 vols. San Francisco: The History Company, 1887.

Going Up the Flume: The Hanging of John Barclay (pages 313–318)

Columbia [CA] Gazette, October 12, 1855.

Columbia [CA] Gazette, October 13, 1855. Quoted in the *Sacramento Daily Union*, October 19, 1855.

Lang, Herbert O. *A History of Tuolumne County, California: Compiled from the Most Authentic Sources*. San Francisco: B. F. Alley, 1882.

Sacramento Daily Union, October 19, 1855.

Tuolumne County Register. Marriage record for John Barclay and Martha Carlos, September 18, 1855, signed by J. A. Copwell, Justice of the Peace.

***"Awful Calamity": The Explosion of the Steamer* Belle** (pages 319–325)

Burnett, Marie. "Belle." From the California State Lands Commission, *Brother Jonathan* Shipwreck Exhibit, http://www.slc.ca.gov/Info/Shipwrecks/Belle.pdf

Holliday, J. S. *Rush for Riches*. Berkeley: University of California Press, 1999.

Hutchings, James Mason. "Editor's Table," *Hutchings' California Illustrated Magazine* 2, no. 11 (May 1858): 524–27.

Maitland [Australia NSW] Mercury and Hunter River General Advertiser, April 17, 1856.

McGowan, Joseph. *Golden Notes* 4, no. 2 (January 1958). Published by the Sacramento County Historical Society.

New York Times, March 14, 1856.

Sacramento Daily Democratic State Journal. "Awful Catastrophe—Explosion of the Steamer Belle—Dreadful Loss of Life." February 6, 1856.

Sacramento Daily Union, February 6, 1856; February 7, 1856; and September 7, 1864.

San Francisco Daily Alta California. "Awful Calamity." February 6, 1856.

San Francisco Daily Alta California. "Explosion of the Steamer Belle." February 8, 1856.

San Francisco Herald, February 20, 1856.

The Territory of Colorado: The Pico Act of 1859 (pages 326–328)

Di Leo, Michael. *Two Californias: The Myths and Realities of a State Divided Against Itself*. Covelo, CA: Island Press, 1983.

Ellison, William Henry. "The Movement for State Division in California, 1849–1860." *Southwestern Historical Quarterly* 17, no. 2 (October 1913): 101–39.

Los Angeles Star. "The Proposed 'Territory of Colorado.'" February 19, 1859.

Pitt, Leonard, and Dale Pitt. *Los Angeles A to Z: An Encyclopedia of the City and County*, 392. Berkeley: University of California Press, 1997.

Statutes of California, Passed at the Tenth Session of the Legislature, 1859, 310–11. Sacramento: John O'Meara, State Printer, 1859.

The Final Accounting: Death in the Goldfields (pages 329–336)

Booth, Edmund. *Edmund Booth, Forty-Niner: The Life Story of a Deaf Pioneer, including Portions of His Autobiographical Notes and Gold Rush Diary, and Selections from Family Letters and Reminiscences*. Stockton, CA: San Joaquin Pioneer and Historical Society, 1953.

Carson, James H. *Early Recollections of the Mines, and a Description of the Great Tulare Valley*. Reprint, Tarrytown, NY: W. Abbatt, 1931. Originally printed as a supplement in the *San Joaquin Republican* (Stockton, CA) in 1852.

Delano, Alonzo. *Life on the Plains and among the Diggings*. Auburn, NY: Miller, Orton and Mulligan, 1856.

Haun, Catherine. "A Woman's Trip Across the Plains, 1849." Original diary at the Huntington Library, San Marino, CA. Reprinted in part in *Women's Diaries of the Westward Journey*, by Lillian Schlissel. New York: Schocken Books, 1992.

Helper, Hinton Rowan. *The Land of Gold: Reality versus Fiction*. Baltimore: Hinton Helper, 1855.

Hutchings, James Mason. "The Miners' Death." *Hutchings' Illustrated California Magazine* 1, no. 7 (June 1857).

Kip, William Ingraham. *The Early Days of My Episcopate*. New York: T. Whittaker, 1892.

Letts, John M. *California Illustrated: Including a Description of the Panama and Nicaragua Routes*. New York: R. T. Young, 1853.

Levy, JoAnn. *They Saw the Elephant: Women in the California Gold Rush*. Norman: University of Oklahoma Press, 1992.

Lienhard, Heinrich. *A Pioneer at Sutter's Fort, 1846–1850: The Adventures of Heinrich Lienhard*. Translated, edited, and annotated by Marguerite Eyer Wilbur from the original German manuscript. Los Angeles: Calafía Society, 1941.

Shaw, William. *Golden Dreams and Waking Realities: Being the Adventures of a Gold-Seeker in California and the Pacific Islands*. London: Smith, Elder and Co., 1851.

Shirley, Dame [Louise Amelia Knapp Smith Clappe]. Dame Shirley to her sister Molly in Massachusetts, September 22, 1851. *The Pioneer* (January 1854–December 1855). Reprinted in *The Shirley Letters: From the California Mines, 1850–1852*. Berkeley: Heyday, 1998, 2001.

Soulé, Frank, John H. Gihon, and James Nisbet. *The Annals of San Francisco*. New York: D. Appleton, 1855.

Steele, John. Journal entry, January 30, 1851. *In Camp and Cabin: Mining Life and Adventure, in California During 1850 and Later.* Lodi, WI: J. Steele, 1901.

Williams, Albert. *A Pioneer Pastorate and Times: Embodying Contemporary Local Transactions and Events*. San Francisco: Wallace and Hassett, printers, 1879.

Woods, Daniel Bates. *Sixteen Months at the Gold Diggings*. New York: Harper and Brothers, 1851.

SOURCES FOR CHAPTER TITLES

"The Wings of the Future" from "To the Rock That Will Be a Cornerstone," in *The Wild God of the World: An Anthology of Robinson Jeffers*. Palo Alto: Stanford University Press, 2003.

"Undisciplined Squads of Emotion" from *Four Quartets, East Coker,* by T. S. Eliot. 1943. Reprint, Boston: Mariner Books, 1968.

"Fearful Blindness" from *California: A Study of American Character: From the Conquest in 1846 to the Second Vigilance Committee in San Francisco,* by Josiah Royce. Boston: Houghton Mifflin, 1886.

"I Know It to Be Nothing Else" from "Personal Account of Gold Discovery," by James Wilson Marshall, in *The Life and Adventures of James W. Marshall, the Discoverer of Gold in California,* by George Frederic Parsons. Sacramento: J. W. Marshall and W. Burke, 1870.

"A Vast Deal of Knavery" from *Wonderful Facts from the Gold Regions: Also, Valuable Information Desirable to Those Who Intend Going to California* by Daniel Walton. Boston: Stacy, Richardson and Co., 1849.

"The El Dorado of Their Most Sanguine Wants" from *Around the Horn in '49: Journal of the Hartford [CT] Union Mining and Trading Company; Containing the Name, Residence and Occupation of Each Member, with Incidents of the Voyage, &c.&c,* by J. Linville Hall. Wethersfield, CT: L. J. Hall, 1898.

"A Perfect Used Up Man," lyric from the Gold Rush song "Life in California," in *The Gold Rush Song Book,* by Eleanora Black and Sidney Robertson. San Francisco: Colt Press, 1940.

"With a Taint of Fraud and a Spice of Comedy" from *History of California,* by Theodore H. Hittell. Vol. 3. San Francisco: N. J. Stone, 1897.

"Death Stared Them Full in the Face" from *Life in the Mountains; or Four Months in the Mines of California,* by Silas Weston. Providence, RI: E. P. Weston, 1854.

"That Blighting Curse" from *California Illustrated: Including a Description of the Panama and Nicaragua Routes*, by John M. Letts. New York: R. T. Young, 1853.

"Plenty of Jabbering and Quarreling and Several Fights" from *The Journals of Alfred Doten, 1849–1903*, by Alfred Doten. 3 vols. Reno: University of Nevada Press, 1973.

"Gambling on One Card the Fruit of His Labor for the Year" from William Peters, letter to Miss Emily Howard, January 1, 1851. Original letter in the California History Section (SMCII, box 13, folder 12), California State Library, Sacramento.

"There Is No Persuasion More Esteemed for Moral Conduct" from Emanuel Linoberg letter to the editor, *American Israelite* (Cincinnati, OH), November 13, 1857.

"Everything Looks Forlorn and Wretched" from *Up and Down California in 1860–1864*, by William Brewer. 1864. Berkeley: University of California Press, 1966.

"A Wind Turned Dark with Burning" from *Fahrenheit 451*, by Ray Bradbury. New York: Ballantine, 1953.

"The Blue Vault of Heaven" from *Old Block's Sketch Book*, by Alonzo Delano. Sacramento: James Anthony, 1856. Reprinted with a foreword by Marguerite Wilbur. Santa Ana, CA: Fine Arts Press, 1947.

"The Devil's Chaos" from *Our New West: Records of Travel between the Mississippi River and the Pacific Ocean*, by Samuel Bowles. Hartford, CT: Hartford Publishing Co., 1869.

"As Huge, to Me, as an Elephant" from *El Campo de los Sonoraenses: Three Years Residence in California: 1849–1852*, by William Perkins. Undated manuscript found in the Stanford University Library, Special Collections, ID M0083.

"A Very Normal Childhood" from *Records of a California Family: Journals and Letters of Lewis C. Gunn and Elizabeth Le Breton Gunn*, edited by Anna Lee Marston. San Diego: 1928.

"The Years Have Been Full of Hardships" from *Luzena Stanley Wilson, '49er: Memories Recalled Years Later for Her Daughter Correnah Wilson Wright*, by Luzena Stanley Wilson. Oakland: Eucalyptus Press, 1937.

"Shadow and Light" from *Shadow and Light: An Autobiography, with Reminiscences of the Last and Present Century*, by Mifflin Wistar Gibbs. Washington, D.C.: 1902.

"A Vast, Glowing Empty Page" from *The Dharma Bums*, by Jack Kerouac, Chapter 21. New York: Viking Press, 1958.

"Emperor of These United States" from a proclamation by Emperor Norton I in the *San Francisco Bulletin*, September 18, 1859.

"Very Little Law of Any Kind" from *Memoirs of Elisha Oscar Crosby: Reminiscences of California and Guatemala from 1849 to 1864*, by Elisha Oscar Crosby. San Marino, CA: Huntington Library, 1945.

"Enjoy It with Luxurious Zest" from "Off to the Mountains," by James Mason Hutchings. *Hutchings' California Magazine* 4, no. 4 (October 1859).

"To the Land of Gold and Wickedness" from *To the Land of Gold and Wickedness: The 1848–54 Diary of Lorena L. Hays*. Edited by Jeanne Hamilton Watson. St. Louis: Patrice Press, 1988.

"The Theatre of Unrest" from *California, In-doors and Out; or, How We Farm, Mine, and Live Generally in the Golden State*, by Eliza W. Farnham. New York: Dix, Edwards and Co., 1856.

"Convulsive Throes and Conflicts of Passion" from "History of Sacramento," by John Morse, in *The Sacramento Directory for the Year 1853–54*, by Samuel Colville. Sacramento: Union Office, 1853. Reprinted with an introduction by Mead B. Kibbey. Sacramento: California State Library Foundation, 1997.

"I Am Joaquín!" from *The Life and Adventures of Joaquin Murieta, the Celebrated California Bandit*, by John Rollin Ridge. San Francisco: W. B. Cooke, 1854.

"A Great Deal of Hard Times" from *Diary of a Forty-Niner, Being the Diary of Joseph Pike of Half Day, Illinois, April 15, 1850, to December 29, 1851*. Original manuscript and typescript in the California History Section, California State Library, Sacramento.

"Guilty of Dust and Sin" from "The Elixir," in *The Temple*, by George Herbert. University of Cambridge: Buck and Daniel, 1633.

"Players and Painted Stages" from "The Circus Animals' Desertion," in *The Collected Poems of W. B. Yeats: Definitive Edition, with the Author's Final Revisions*. New York: Macmillan, 1956.

"So Wonderful, So Dangerous, So Magnificent a Chaos" from *Eldorado: Adventures in the Path of Empire*, by Bayard Taylor. New York: G. P. Putnam, 1850.

"The Legislature of a Thousand Drinks" was a comical reference to California's first state legislature and is attributed to Thomas Jefferson Green.

"An Act for the Better Regulation of the Mines and the Government of Foreign Miners," from *Statutes of California Passed at the First Session of the Legislature, December 15, 1849, to April 22, 1850, at the Pueblo de San José*, Chapter 97, 221–23. San José: J. Winchester, State Printer, 1850.

"We Are Not the Degraded Race You Would Make Us" from "To His Excellency Governor Bigler," letter from Noman Asing, *San Francisco Daily Alta California*, May 5, 1852.

"The Theatre of Scenes which We Hope Will Never Be Repeated" from *Sacramento News-Letter*, August 30, 1851.

"Going Up the Flume" was a nineteenth-century slang term for death. In Barclay's case it was literally true.

"Awful Calamity" from *San Francisco Daily Alta California*, February 6, 1856.

"The Territory of Colorado" from "An Act Granting the Consent of the Legislature to the Formation of a Different Government for the Southern Counties of This State." Also known as the Pico Act of 1859, it was approved by the California State Legislature on April 18, 1859.

"The Final Accounting" from the Latin phrase "Mors ultima ratio," or "Death is the final accounting."

ACKNOWLEDGMENTS

The image of the solitary writer toiling away in the flickering candlelight with a quill, scratching out a finished manuscript, ready for publication, is a romantic, evocative vision. In reality, from the earliest days of publishing, the production of a book is very much a team effort. My team was exemplary and deserves much praise and credit for *Gold Rush Stories: 49 Tales of Seekers, Scoundrels, Loss, and Luck.*

My heartfelt appreciation and greatest respect to my colleague and mentor Malcolm Margolin, founder and former publisher of Heyday. This book was begun during his reign and benefitted greatly from his sage advice. My best wishes to Malcolm for a long and happy retirement. Thank you, my friend.

My warmest admiration is extended to the extraordinary crew at Heyday—editorial director Gayle Wattawa; editors Eve Bachrach and Lisa K. Marietta; outreach director Lillian Fleer; marketing director Mariko Conner; designer Ashley Ingram; art director Diane Lee; and all the remarkable personnel that comprise the amazing Team Heyday.

Extra-special thanks to my friend Gary Kurutz, Director of Special Collections at the California State Library, for his thoughtful and elegant introduction to *Gold Rush Stories.* I spent many hours in the library's California History Section researching the California Gold Rush and being constantly astonished by the depth and splendor of the collections. I was blessed to enjoy the extensive guidance of Gary and the incredible staff of the State Library in developing and writing this book. I am eternally grateful for their help and encouragement. They truly went the extra mile.

My thanks and deepest appreciation to my Sierra College Press family, led by my friend, editor-in-chief Joe Medeiros. Thank you to all for your endless support and friendship. I could not have done this without you.

And, finally, my endless gratitude to the two people who spurred my abiding fascination with all things Gold Rush: my father, Howard Noy, and my grandfather, Thomas Noy. Both were Cornish old-school gold miners who labored underground for years in search of the elusive golden quarry. From their mining experience, they passed on to me an important lesson that I have tried to follow, to the best of my ability, throughout my life: namely, keep your head down, keep digging, and all the hard work will pay off. Thanks, Dad and Grandpa. *Gold Rush Stories* is respectfully and lovingly dedicated to you.

ABOUT THE AUTHOR

A Sierra Nevada native, Gary Noy has taught history at Sierra Community College in Rocklin, California, since 1987. A graduate of UC Berkeley and CSU Sacramento, he is the founder and former director of the Sierra College Center for Sierra Nevada Studies and the editor-in-chief emeritus of the Sierra College Press. In 2006, the Oregon-California Trails Association (OCTA), a national historical society, selected Noy as Educator of the Year. He is the author of *Distant Horizon: Documents from the Nineteenth-Century American West* (1999), the coeditor, with Rick Heide, of *The Illuminated Landscape: A Sierra Nevada Anthology* (2010), and the author of *Sierra Stories: Tales of Dreamers, Schemers, Bigots, and Rogues* (2014). In 2016, *Sierra Stories* received the Gold Medal for Best Regional Non-Fiction from the Next Generation Indie Book Awards.

Photo by Mike Price. Courtesy of Sierra College Press.

ABOUT THE SIERRA COLLEGE PRESS

In 2002, the Sierra College Press was formed to publish *Standing Guard: Telling Our Stories* as part of the Standing Guard Project's examination of Japanese American internment during World War II. Since then the Sierra College Press has grown into the nation's first academic press operated by a community college.

The mission of the Sierra College Press is to inform and inspire scholars, students, and general readers by disseminating ideas, knowledge, and academic scholarship of value concerning the Sierra Nevada region. The Sierra College Press endeavors to reach beyond the library, laboratory, and classroom to promote and examine this unique geography.

For more information, please visit www.sierracollege.edu/press.

Special thanks to Sierra College Friends of the Library, a major financial supporter.

ABOUT HEYDAY

Heyday is an independent, nonprofit publisher and unique cultural institution. We promote widespread awareness and celebration of California's many cultures, landscapes, and boundary-breaking ideas. Through our well-crafted books, public events, and innovative outreach programs we are building a vibrant community of readers, writers, and thinkers.

THANK YOU

It takes the collective effort of many to create a thriving literary culture. We are thankful to all the thoughtful people we have the privilege to engage with. Cheers to our writers, artists, editors, storytellers, designers, printers, bookstores, critics, cultural organizations, readers, and book lovers everywhere!

We are especially grateful for the generous funding we've received for our publications and programs during the past year from foundations and hundreds of individual donors. Major supporters include:

Anonymous; Arkay Foundation; Richard and Rickie Ann Baum; Randy Bayard; Jean and Fred Berensmeier; Edwin Blue; Jamie and Philip Bowles; John Briscoe; California Humanities; John and Nancy Cassidy; Graham Chisholm; The Christensen Fund; Jon Christensen; Steve Costa and Kate Levinson; Lawrence Crooks; Nik Dehejia; Topher Delaney; Chris Desser and Kirk Marckwald; Frances Dinkelspiel and Gary Wayne; The Roy and Patricia Disney Family Foundation; Tim Disney; Marilee Enge and George Frost; Richard and Gretchen Evans; The Stoddard Charitable Trust; John Gage and Linda Schacht; Patrick Golden and Susan Overhauser; Whitney Green; Penelope Hlavac; Charles and Sandra Hobson; Nettie Hoge; Donna Ewald Huggins; JiJi Foundation; Claudia Jurmain; Kalliopeia Foundation; Marty Krasney; Abigail Kreiss; Guy Lampard and Suzanne Badenhoop; The Campbell Foundation; Thomas Lockard and Alix Marduel; David Loeb; Judith

Lowry-Croul and Brad Croul; William, Karen, and John McClung; Nion McEvoy and Leslie Berriman, in honor of Malcolm Margolin; Mark Murphy; Richard Nagler; National Wildlife Federation; The Nature Conservancy; Steven Nightingale and Lucy Blake; Eddie Orton; Julie and Will Parish; Ronald Parker; The Ralph M. Parsons Foundation; James and Caren Quay; Melissa Riley; The San Francisco Foundation; San Manuel Band of Mission Indians; Greg Sarris; Ron Shoop; Roselyne Swig; Tappan Foundation; Thendara Foundation; Michael and Shirley Traynor; Stanley Smith Horticultural Trust; The Roger J. and Madeleine Traynor Foundation; Al and Ann Wasserman; Sherry Wasserman and Clayton F. Johnson; Lucinda Watson; Valerie Whitworth and Michael Barbour; Cole Wilbur; Peter Wiley and Valerie Barth; and the Yocha Dehe Wintun Nation.

BOARD OF DIRECTORS

GETTING INVOLVED